Playful Frames

TMI TECHNIQUES of the MOVING IMAGE

Volumes in the Techniques of the Moving Image series explore the relationship between what we see onscreen and the technical achievements undertaken in filmmaking to make this possible. Books explore some defined aspect of cinema—work from a particular era, work in a particular genre, work by a particular filmmaker or team, work from a particular studio, or work on a particular theme—in light of some technique and/or technical achievement, such as cinematography, direction, acting, lighting, costuming, set design, legal arrangements, agenting, scripting, sound design and recording, and sound or picture editing. Historical and social background contextualize the subject of each volume.

Murray Pomerance
Series Editor

Jay Beck, *Designing Sound: Audiovisual Aesthetics in 1970s American Cinema*
Lisa Bode, *Making Believe: Screen Performance and Special Effects in Popular Cinema*
Wheeler Winston Dixon, *Death of the Moguls: The End of Classical Hollywood*
Nick Hall, *The Zoom: Drama at the Touch of a Lever*
Andrea J. Kelley, *Soundies Jukebox Films and the Shift to Small-Screen Culture*
Adrienne L. McLean, *All for Beauty: Makeup and Hairdressing in Hollywood's Studio Era*
R. Barton Palmer, *Shot on Location: Postwar American Cinema and the Exploration of Real Place*
Murray Pomerance, *The Eyes Have It: Cinema and the Reality Effect*
Steven Rybin, *Playful Frames: Styles of Widescreen Cinema*
Colin Williamson, *Hidden in Plain Sight: An Archaeology of Magic and the Cinema*
Joshua Yumibe, *Moving Color: Early Film, Mass Culture, Modernism*

Playful Frames

Styles of Widescreen Cinema

STEVEN RYBIN

Rutgers University Press
New Brunswick, Camden, and Newark, New Jersey
London and Oxford

Rutgers University Press is a department of Rutgers, The State University of New Jersey, one of the leading public research universities in the nation. By publishing worldwide, it furthers the University's mission of dedication to excellence in teaching, scholarship, research, and clinical care.

Library of Congress Cataloging-in-Publication Data

Names: Rybin, Steven, 1979– author.
Title: Playful frames : styles of widescreen cinema / Steven Rybin.
Description: New Brunswick : Rutgers University Press, [2023] | Includes bibliographical references.
Identifiers: LCCN 2023008684 | ISBN 9781978815957 (cloth) | ISBN 9781978815940 (paperback) | ISBN 9781978815964 (epub) | ISBN 9781978815988 (pdf)
Subjects: LCSH: Wide-screen processes (Cinematography)—Case studies | Motion picture producers and directors—United States—Case studies.
Classification: LCC TR855 .R93 2023 | DDC 777—dc23/eng/20230415
LC record available at https://lccn.loc.gov/2023008684

A British Cataloging-in-Publication record for this book is available from the British Library.

Copyright © 2024 by Steven Rybin
All rights reserved

No part of this book may be reproduced or utilized in any form or by any means, electronic or mechanical, or by any information storage and retrieval system, without written permission from the publisher. Please contact Rutgers University Press, 106 Somerset Street, New Brunswick, NJ 08901. The only exception to this prohibition is "fair use" as defined by U.S. copyright law.

References to internet websites (URLs) were accurate at the time of writing. Neither the author nor Rutgers University Press is responsible for URLs that may have expired or changed since the manuscript was prepared.

∞ The paper used in this publication meets the requirements of the American National Standard for Information Sciences—Permanence of Paper for Printed Library Materials, ANSI Z39.48–1992.

rutgersuniversitypress.org

For Jessica

Contents

Playful Frames

Introduction

A Scope Quartet

The lights darken; the curtains part. Receding fabric reveals a vast screen, a blank canvas. An approximate square metamorphoses into an increasingly impressive rectangle, an unveiling of cinematic potential preceding the projection of a widescreen movie. On this capacious screen, any image might find vibrant life. But the very thought of a wide, big screen will conjure for many an ironically narrow range of subject matter: the action movie, the fantasy film, the war epic. CinemaScope, Panavision, and other anamorphic widescreen technologies are often taken as ideal means to enable immersion into sublime landscapes, overwhelming special effects, and thrilling derring-do. There are historical reasons for these still-prevalent associations of widescreen cinema with absorption and spectacle. In the years preceding the introduction of CinemaScope by the commercial film industry in the United States in 1953, Hollywood enjoyed a temporary commercial boom via 3D technology. As David Bordwell reminds us, "the idea of immersing the audience," part of the initial ballyhoo of the short-lived 3D wave in the early 1950s, "was promoted by the backers of widescreen systems" (*Poetics of Cinema* 286). The one-sheet poster advertising Jean Negulesco's *How to Marry a Millionaire* (1953) proudly declared, beneath the curvature of text exclaiming "CinemaScope" that seemed to gesture beyond the borders of the poster toward the beholder of the image, "You See It without Glasses!" John Belton has described this spectacular address to the widescreen viewer in the 1950s as one that solicits physical, immersive participation (194–195). This early association between widescreen cinema and sensorial immersion shaped the subject matter and decor of more than a few early CinemaScope and Panavision films: the biblical marvel *The Robe* (Henry King, 1953; shot in CinemaScope, aspect ratio 2.55:1); the adventure *Around the World in 80 Days* (Michael Anderson and John Farrow, 1956; shot in Todd-AO, 2.20:1); and the grand pyramids of Howard Hawks's *Land of the Pharaohs* (1955; shot in CinemaScope, 2.55:1), to cite just a few examples. (See Figure I.1 for an illustration of some, though not all, cinematic aspect ratios; the territory marked "2.35:1" will be, more or less, the province of this book; and see David Pratt, "Widescreen Box Office Performance to 1959," for a comprehensive look at the most popular widescreen films during the post-1952, pre-1960 period, and one that tends to confirm the notion that the commercial appeal of widescreen is rooted in spectacular genres.)

This "initial ballyhoo" eventually gave way to the standardization of widescreen technology in Hollywood, as CinemaScope, Panavision, and their technological brethren became invisible, increasingly conventional vehicles for film narrative. As the years went on, "Scope" (a shortened word used by many to denote movies shot in a ratio of 2.35:1 or thereabouts, even after CinemaScope's demise) was domesticated both by the norms of standard industry practice, which began to flexibly regulate spectacular excess within the context of traditional narrative norms (see Belton 199), and by the rise, in the 1970s and 1980s, of multiplex and shopping mall cinemas. Such situations of viewing, in so-called shoebox auditoriums, at times even situated the approximately 2.35:1 widescreen image as a reduction of a screen with a native shape closer to 1.85:1. In these exhibition contexts, the curtains masking the screen—if the auditorium in question even had curtains—do not part from the center but are, rather, adjusted horizontally, an effect that tends to shape widescreen as experiential diminution rather than expansion. This standardization, although reductive of widescreen's grandeur, reminds us that the technology was never intrinsically beholden to the spectacular; any subject matter, including intimate, grounded forms of realism (see Belton 201–203), could find a home in widescreen, with directors deploying fine-grained ways to activate areas of the wide image.

Nevertheless, and despite the various kinds of widescreen films created in the decades after widescreen's adoption in Hollywood, the idea that the format is best suited for immersive spectacle remains dominant in the discourse of many viewers, some scholars, and more than a few filmmakers. It is not difficult to find a kind of technological determinism, or essentialism, in chatter about widescreen even in our present-day digital age. "Movies shot in widescreen feel cinematic," one contemporary website dedicated to teaching digital video production informs us. "They *innately* make things feel intense and epic" (Berry; italics mine). Even academic discourse occasionally suggests that widescreen technology is essentially suited for immersive effects. Scholars writing of other subjects, when noting in passing the effects of widescreen, remark on "the immersive address of wide-screen" (Keller 136)—*the* immersive address, as if none other were available; and on how the "orchestration of scale, cinematography, and kinesthetic stimulus" in widescreen films generates "an *immersive* viewing experience" (Taylor 21; italics mine). One article on the psychoanalytic effects of widescreen, or

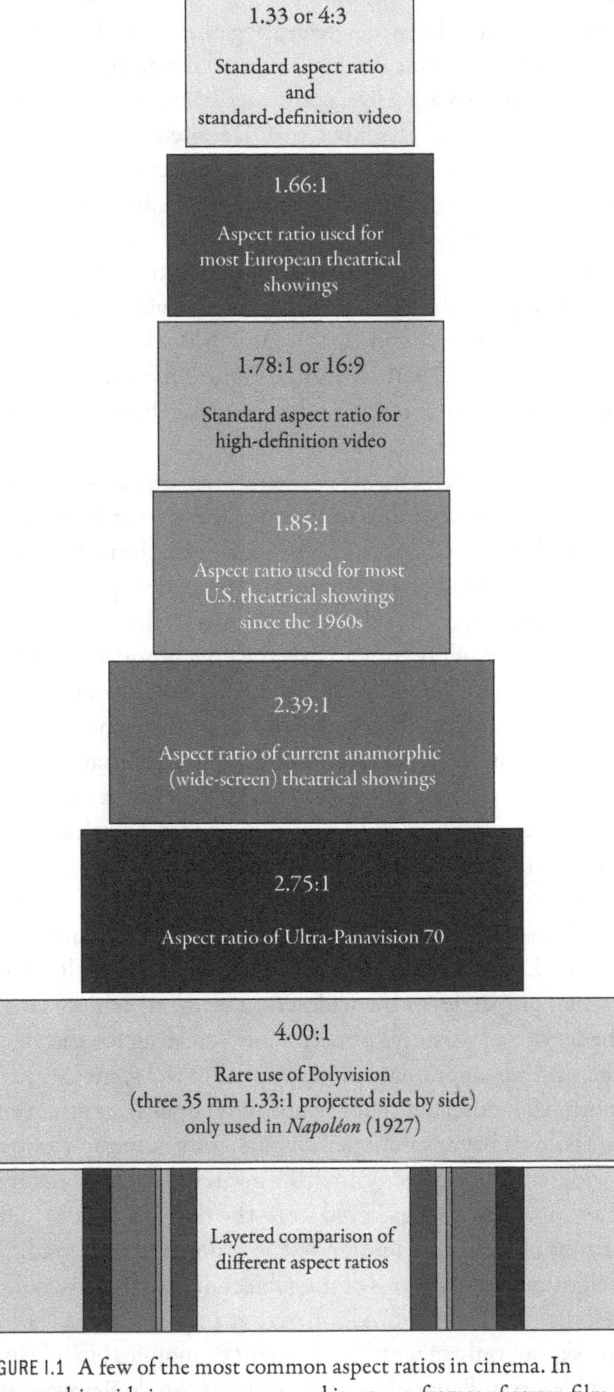

FIGURE I.1 A few of the most common aspect ratios in cinema. In anamorphic wide imagery, a squeezed image on frames of 35mm film is unsqueezed by the projector's anamorphic lens during projection. (Source: Wikimedia Commons)

what the author takes as "immersion cinema," removes aesthetics and social critique from the potential of widescreen altogether, remarking that the medium reflexively "emphasizes technical achievement to the detriment of social or artistic relevance and embeds a passive, consumerist ideology within the spaces of contemporary moviegoing" (Recuber 316). Despite the variety of artistic practice constituting widescreen cinema, not much seems to have changed since Charles Barr observed, in his pioneering essay on CinemaScope in 1963, that most critics at the time "condemned" Scope "from the start as a medium for anything other than the spectacular and the trivial" (5). As David Bordwell observes in an essay on the relationship between widescreen technology and mise-en-scène criticism written in 1985, much work remains to be done on the range of styles, subject matter, and aesthetic effects widescreen technology has historically accommodated ("Widescreen Aesthetics and Mise en Scène Criticism").

The range of widescreen cinema aesthetics remains generally underappreciated in writings on film style and technique. The format is often mentioned in passing in critical studies of directors who work with the technology, but auteur studies rarely take techniques of composing the anamorphic image as a central focus. The orchestration of the relationship between the widescreen aspect ratio and the subject matter of the film is nevertheless a promising path in the appreciation of film art and directorial achievement. Lisa Dombrowski offers a rare example of close attention to the moment-by-moment aesthetic economy of what I take to be alternative widescreen filmmaking—films that express an interest in visions other than the sensorially immersive. She explores stylistic experimentations in widescreen in black and white, low-budget American films of the late 1950s, such as William F. Claxton's *Young and Dangerous* (1957) and Samuel Fuller's *Forty Guns* (1957) (65–66), works that are the playful Hollywood equivalents of Scope films made in the French and Japanese New Waves during roughly the same period. Dombrowski, for example, explores how Fuller, in *Forty Guns*, "uses the formal properties of the widescreen frame to achieve lateral and axial depth in the service of narrative punctuation," creating for the viewer "a recognition of the shifting emotional landscape within the scene" (65). This sense of shifting emotional landscapes that Dombrowski draws our attention to in Fuller's screen art holds at bay any notion of an inherently immersive, one-dimensional address in widescreen technology, inviting awareness of both the frame and the textured emotional resonances keyed in to the development of subject matter. For Dombrowski, aesthetic attention to the frame enables a gradated appreciation of the emotional dimensions of the subject matter as those dimensions take on shape, moment by moment. Alternative examples such as those pointed to by Dombrowski not only suggest the different economies of scale in which CinemaScope, widescreen Panavision, and similar technologies were and continue to be deployed but also point to the existence of a corpus of relatively idiosyncratic widescreen visions and styles, many of which offer alternatives to the commercial film industry's overvaluation of immersion, psychological suture and identification, and narrative consumption.

Filmmakers who approach the wide screen as an opportunity for fresh incarnations of screen art reward our close attention to techniques of screen composition and the various perceived relationships possible between aspect ratio and subject matter. As John Gibbs and Douglas Pye write in an essay on the use of widescreen by Otto Preminger and Sam Peckinpah, "Our interest here is in aspect ratio as one of the areas of choice in their work that affect how the films' worlds are presented to us and that therefore help shape the relationship between spectator and fiction" (71). *Playful Frames*, following from this approach, looks closely at the work of four accomplished filmmakers who often are overlooked in the scholarly literature on widescreen cinema: Jean Negulesco, Blake Edwards, Robert Altman, and John Carpenter. These four auteurs have created distinctive, idiosyncratic styles of widescreen filmmaking, as they do so offering us inventive and expansive visions while also, at times, critiquing the very idea of sensorial immersion that so often defines the nature of the typical commercial Hollywood movie. In approaching their films, I work with a loose definition of widescreen that refers to films with aspect ratios of approximately 2.35:1. In studying the films of these four directors, *Playful Frames* moves across a swath of films, styles, and genres, beginning in 1953 and ending in the first decade of the new century (although not exactly in that order). In doing so it offers a quartet of essays on the aesthetic effects of widescreen composition across shifting historical backgrounds, in various genres, and in relation to different types of subject matter.

In the chapters that make up the book, I explore how Negulesco, Edwards, Altman, and Carpenter creatively—and at times slyly and mischievously—use the widescreen image in playful, rather than norm-defining or "innovatively path-breaking," ways. Their films generate a sense of widescreen technology as a path toward alternative cinematic aesthetics, work that encourages an awareness of the artistic construction of the image in and against the historical period in which the films were made. Playful widescreen frames are, for me, not always already "immersive" but are rather intellectually engaging and emotionally surprising, inviting the viewer to explore the perceptually creative, underdetermined, and generative spaces possible in the art of cinema. Further, these four directors make widescreen films for reasons that run counter to the dominant discourses about the technology in the industry (sometimes with a sharp pang of late-career bitterness about the fate of art in a world dominated by commerce, particularly in Edwards and Altman). These alternative sensibilities emerged even as all four filmmakers navigated commercial and genre imperatives and even as each produced a string of successful entertainments using the techniques made possible by widescreen technologies.

A few additional reasons justify the selection of these four filmmakers, beyond my own preferences. For one, the two most important book-length monographs in English on the technology, aesthetics, and history of widescreen cinema—John Belton's *Widescreen Cinema* (1992) and Harper Cossar's *Letterboxed: The Evolution of Widescreen Cinema* (2011)—either ignore these directors entirely (Altman, Carpenter, Edwards) or mention them only perfunctorily (Negulesco, who is usually only mentioned because he directed the first film made—and the second released to the public—in CinemaScope, *How to Marry a Millionaire*, 1953). But this quartet of directors ran against the grain of the industry's tendency to position widescreen, and screen entertainment generally, as immersive, spectacular entertainment. Of course, historical studies of widescreen cinema have themselves gone beyond immersion and spectacle, looking closely at how the format has been thoughtfully used in the development of narrative norms: Belton and Cossar, for example, examine exemplary artists of the wide image such as Anthony Mann, Vincente Minnelli, Otto Preminger, Nicholas Ray, and Frank Tashlin. By expanding the poetics of widescreen cinema technique to include Negulesco, Edwards, Altman, and Carpenter, and by stretching my priorities beyond the historical establishment of "norms" of filmmaking, *Playful Frames* considers the creative use of widescreen in other film genres and authorial sensibilities, studying the ways these directors created idiosyncratic and variously textured relationships between aspect ratio and subject matter.

Despite their individually distinctive styles, this quartet of directors are for me united by a shared sensibility. Negulesco, Edwards, Altman, and Carpenter share a playful aesthetic approach, one that mischievously circumvents expectation and convention—sometimes in blunt or provocative ways, at other times in subtle and elegant ones—even as each filmmaker works in popular genres. In ways inspired by the relationship between film art and the pleasures of play, this study sees techniques of widescreen composition, and the ensuing perception of techniques, as potentially ludic and puckishly disruptive crafts. Widescreen, as it is practiced by these four directors, is an artistic practice that goes beyond the norms and expectations of narrower, replicative ways of seeing, taking us beyond spectacle and expectation. Ariel Rogers has shown in her study of embodiment in the context of emerging cinematic technologies that the new scale and shape of the widescreen format in the 1950s initially "elicited a feeling of immersion by downplaying the viewer's awareness of the limits of the frame and, as a result, obscuring the barrier between viewer and spectacle" (30). I agree that this has been true of much widescreen practice, then and now. But it is my contention that the four filmmakers under study here, rather than complying with the industry's imperative to create work that would lure the viewer to forget the presence of the frame, sought instead to make the viewer saliently aware of the constructed nature of the wide image in relation to, rather than as an invisible canvas for, subject matter. Such awareness encourages us to analyze their films as a set of aesthetic objects enabled by a careful attention to techniques of composition rather than as further evidence that widescreen technology somehow inherently encourages absorption and that it is domesticated by classical narrative norms that repress distinctive and unruly stylistic gesture. If this book seeks to draw attention to the relationship between screen composition and subject matter, it is in response to the ways these four filmmakers have already enabled playful and artistic attentiveness in their work.

A naysayer could argue all I am after here is a sense of critical distance and viewer awareness that can be brought to any film, regardless of director or genre or technological format. We can always become aware of the frame in cinema, after all, warding off immersive enticements the culture industry dangles in front of us perpetually (in doing so dutifully distinguishing ourselves from that bogeyman of film studies, the "naive viewer"). But aesthetic distance and aesthetics more generally are not identical to the concept of "critical awareness," and often, in cinema studies and other disciplines, being critically distanced, ethically aware, or theoretically sophisticated offers no sure path to substantive and creative discussion of the aesthetic implications of film technique. Further, these four directors draw our attention to the aesthetic fact of the frame not through strategies of overt or salient critical distancing, as Jean-Luc Godard does in his modernist widescreen experiments, such as *Pierrot le fou* (1965) and *2 or 3 Things I Know about Her* (1967). Instead, my filmmakers establish subtle self-consciousness and elegant reflexivity in their use of the widescreen frame from within the context of Hollywood commercial filmmaking. Arguably, a playful aesthetic distance in their work becomes even more perceptible today when their films are screened on DVD or Blu-ray, given how the black bars preserving the frame of 2.35:1 films on television sets call attention to the particular shape of the frame against the standardized 16:9 HDTV aspect ratio. In an earlier era, with its standard-definition 1.33:1 television sets prevalent from the 1950s through the 1990s, rectangular wide frames were similarly letterboxed within a squarish televisual image. (These contexts of home viewing could be understood as transcriptions of the aforementioned tendency of some movie auditoriums to situate the 2.35:1 frame as a reduction of a native 1.85:1 canvas.) But this penchant for a playful and intellectually interactive style was always salient in the varied ways these four filmmakers approached techniques of screen composition.

Balancing Immersion and Aesthetics

André Bazin once complained about problems in film presentation in certain Parisian cinemas, theaters equipped with the technology necessary to project CinemaScope prints but lacking in the expertise and craft necessary to exhibit them properly. His comments strikingly prefigure certain inconsistencies and contingencies in the presentation of films in cinemas and on high-definition television sets in the twenty-first century. In a 1953 essay ("Massacre in CinemaScope") on the future of CinemaScope in the film industry, Bazin made the following observation about the unfortunate contingencies of theatrical widescreen experience:

> Another annoyance occurs quite frequently in widescreen cinemas. When a film in a smaller format is projected and doesn't make use of the whole screen, it often doesn't occur to the projectionist to mask the white margins with a black curtain or an adjustable frame, so the fringes of the image spread out flaccidly into a grayish zone that diminishes the contrast between the real universe and the film. The window (or masking) effect essential to the cinematic illusion is disrupted by the evidence of a vague stain of light projected on a white surface. Moreover, when this concave surface (again, this is an improvement) is vertically striated with visible seams in all the illuminated parts of the image, the reality of the screen's fabric has succeeded in destroying the cinematic illusion. This kind of screen is known as the "miracle screen," apparently because it brings about the miracle of preventing us from believing in cinema. (265)

Too often characterized as a humanist theorist whose main preoccupation was with the gradual achievement of an inevitable, all-encompassing realism through each gradual stage in the evolution of film technology, Bazin was a sensitive critic of films—and, as the preceding quote suggests, at times a quite irritable critic of the exhibition of films—who was sharply attuned to aesthetics and to the various technological, social, interpersonal, and economic contingencies of film presentation. Here he expresses a thoroughly justified testiness at the inability of certain projectionists to correctly present widescreen films through a correct masking of the screen. His comments on the "grayish zone" that obtains when the exhibitor fails to correctly mask the projected film image prefigure, decades ahead of the fact, the inconsistent presentations of widescreen films in U.S. cinemas that, in the early 2010s, began to replace 35mm film projectors and older film screens with new digital cinema projection technology, as well as the various contingencies involved with viewing widescreen compositions at home on televisions, tablets, and other devices.

In today's landscape of multiplex, art-house, and repertory film exhibition, images crafted in an aspect ratio of 2.35:1 (or wider)—that is, images that are approximately 2.35 times wider than they are tall—are often presented as a narrowing of a standardized, fixed projection screen. Many multiplex cinema chains in the United States at the time of my writing are adopting "fixed" aspect ratios in their auditoriums—in my moviegoing experience, most often "fixed" at something like 16:9 or 2:1, although I have also seen fixed theatrical aspect ratios at approximately 2.35:1 (screens that require the projection of vertical black bars on the side of the image when 2:1, 1.85:1, or narrower films are screened in such auditoriums). In a "fixed" ratio theatrical screen narrower than 2.35:1, the image is adjusted to accommodate 2.35:1 films not through the masking of the image with the deep black of a curtain's fabric adjusted to match the particular aspect ratio of the film currently booked in the auditorium, but more frequently (and not unlike what we see when watching widescreen films on televisions) with the digital projection of black bars onto the screen itself, bars that are never dark enough and so never quite disappear during the presentation. In terms of home viewing, the watching of films on iPads and other mobile viewing devices reduces the perceived or potential grandeur or scale of widescreen even further. These "black bars" that accompany a widescreen film on such devices are never quite as black as they should be, unlike the adjustable fabric that shapes our sense of a frame in a carefully designed cinema auditorium; reminiscent of Bazin's "grayish zone," the onscreen letterboxing of widescreen films (in both home viewing and fixed-ratio theatrical exhibition contexts) is itself a visible color, less the borders of a masked frame and more a "vague stain of light" that becomes part of the film rather than a cordoning off of our framed experience of it.

Bazin expresses a passion for proper masking of the widescreen film image precisely so that the "illusion" of cinema may be preserved. But we should be careful in assuming what "illusion" means in Bazin's thinking. For Bazin, the borders of the frame never quite disappear, even in the ideal exhibition scenario. Bazin of course did not proselytize for a thoughtless or overwhelmed immersion in the illusion of reality that the film industry understood CinemaScope to enable, and his writings do not encourage us to lose awareness of the aesthetic contours of the film frame, the sense of a film object created by the four borders surrounding the image. His writings on cinema encourage a careful balance between our desire to recognize each stage in the evolution of film technology as another gesture toward the representation of "reality" and a thoughtful awareness of the artificiality, indeed the aesthetic, of the wide image.

It is not a surprise that Bazin and his closest intellectual colleagues, the future filmmakers of the French New Wave who wrote in *Cahiers du Cinéma*, should have been fascinated by emerging CinemaScope technology in the 1950s. During this period, widescreen promised fresh possibilities for both filmmakers and viewers, enabling striking variations on the cinematic signatures of auteurs and, in turn, engaged forms of film criticism written by those who loved the cinema and who were hungry for perceptually creative ways to think about it. These critics also had cultural and historical stakes in the arrival of widescreen as a viable commercial and industrial practice: as Catherine Jones and Richard Neupert point out in the translators' introduction to a 1985 dossier on widescreen that included new translations of essays on CinemaScope by André Bazin, "The *Cahiers* critics were intrigued to see how this originally French concept of the wide screen, springing from Henri Chrétien's panoramic hypergonar lens of the twenties, would be used by Hollywood" (see Bazin, "Three Essays on Widescreen" 8).

However, Bazin's own writings during this period of widescreen's emergence in Hollywood are less interested in the French origins of the technology and more in speculation about the implications CinemaScope will have for practitioners of film realism. Near the end of one of his most important pieces on CinemaScope, "Will CinemaScope Save the Cinema?" (included in the 1985 republication of three of his essays in *The Velvet Light Trap*, "Three Essays on Widescreen"), Bazin makes very clear the potential of widescreen for filmmakers invested in cinematic realism:

> Widescreen cannot help but hasten what we love within the most modern tendencies of cinema: the shedding of all artifice extrinsic to the image's content itself, of all expressionism of time and space. Cinema will further distinguish itself from music and painting. It will draw nearer to its profound vocation, which is to show before expressing, or more precisely, to express by the evidence of the real, that is to say, not so much to signify as to reveal. (14)

In such passages Bazin aligns widescreen with a larger notion of cinema as an artistic practice meant to guide the viewer's eye toward a more expansive and more intensely ambiguous experience of "reality." Bazin is intrigued by how CinemaScope might function as the next evolutionary step in a gradual progression of the asymptotic line cinema forms as it gradually (but, as Bazin knew, never completely) approaches the axis of a full representation of the "real" world. Such emphasis rests on the idea that widescreen complements the depth of focus already achieved in experiments in 1930s and 1940s cinematography with an expansion of the frame's potential for providing lateral views more closely representing how a human eye sees the world.

There is no question that Bazin's writings on CinemaScope are interested in the possibility of the technology enabling a gradually fuller plenitude of cinematic realism. But Bazin does not essentialize this property of the wide image. His essays are also replete with intriguing ambiguities and self-conscious musings on the wide frame. As David Bordwell points out, Bazin acknowledged the cinema gets closer to a representation of reality with each significant technological advancement but, crucially, always with the addition of fresh forms of artifice that throw erstwhile aesthetic aspects of the cinema once more into relief (see "Widescreen Aesthetics"). Bazin also acknowledges that cinema is framed against a larger historical and economic backdrop that situates each film as a particular aesthetic object in the time and place of its production. In another essay, for example, he remarks about how "the film emulsion necessarily fixes *the artwork* in a particular historical and social context" ("On Realism" 5; italics mine) and of the individual film performer's importance to our perception of the drama, the actor "interpreting society through their slightest gestures, the way they walk or smile" ("On Realism" 6). By drawing our attention to individual manifestations of gesture and movement, Bazin reminds us that the reality toward which cinema moves is itself never a stable thing. CinemaScope, like all forms of cinema before and after it, is subject to the contingencies and ambiguities inherent in reality itself, to the various contexts and individual expressions that make up an ongoingly transformative and ongoingly transformed reality of which the technology is a part.

In "Will CinemaScope Save the Cinema?" Bazin also acknowledges—and discusses at length—the ways in which widescreen is utterly dependent on the unpredictable flux and flow of both audience taste and Hollywood's economic bottom line:

> Obviously cinema is not lacking for inventors and even geniuses that have made it advance as irrefutably as the traditional arts. However, there is no reason to be scandalized that these artists are generally subject to the demands of mass consumption. These constraints have also contributed to the grandeur of cinema and it has drawn very positive aesthetic profits from them. . . . But normal advances, although difficult, are completely at the mercy of the technical upheavals which can interrupt their course for purely economic reasons. Thus, the silent film had reached an admirable point of perfection when sound came to throw everything into question again. (Bazin, "Three Essays on Widescreen," 11)

Elsewhere in the CinemaScope essay, Bazin holds onto a notion that there exists "the destiny of cinema as art" (12), its ineluctable fate (at least in its chemical, photographic, predigital incarnation) to represent the ambiguity of reality regardless of the intentions of the director and in spite of technological innovation's occasional disruption of previously settled norms and practices. But this notion nevertheless does not prevent Bazin from returning, time and time again, precisely *to* those filmmakers, and to descriptions and analyses of the varied ways they engage with their ever-changing medium. While remaining skeptical of the cult of personality generated by the *politique des auteurs* so fervidly endorsed by his younger colleagues at *Cahiers*, Bazin nevertheless looked closely at the practice of individual directors, and the complexity of their art throughout his writings is a central part of the ontological and social evolution of cinema. If the movement toward greater and greater realism is key for Bazin, the widescreen format is still "superior for 'fictional' mise en scène. . . . The wide screen. . . . allows [filmmakers], apart from the editing process, a freer play" (qtd. in Gibbs 78). For Bazin, widescreen's gesture toward a fuller and fuller "reality" never simply becomes the "immersion" peddled by the commercial film industry. The "freer play" with artistic staging and form is always in complex aesthetic dialogue with a "reality" that is itself always variously unfurling.

Because, for Bazin, this freer play exists in relation to economic, industrial, and social matters, it becomes clear that the "reality" toward which widescreen filmmaking in the 1950s is moving is an ongoingly unpredictable one, an aesthetic frame thrown around various economic, social, and technological contingencies that filmmakers also grapple with in the subject matter of the films themselves.

This reality itself is inflected by the contingencies of individual response and the shifting determinants of a capitalistic economic system dependent on the creation of endlessly variable and consumable experiences. The larger socioeconomic context of film viewing, filmmaking, and technology is part of the ambiguous and contingent social reality CinemaScope (and later iterations of the widescreen aspect ratio) artistically explores and refigures.

Polymorphous Widescreen

My interests in this book lie in the styles and themes of four filmmakers who developed forms of alternative, and frequently anti-immersive, approaches to widescreen aesthetics. The following chapters analyze screen composition techniques by Negulesco, Edwards, Altman, and Carpenter (noting, throughout, the collaborative relationships with the various cinematographers who help shape their compositions). Through a careful consideration of films by this quartet of auteurs, I hope to arrive at a larger holistic sense of what a playful, alternative widescreen cinema might be. My writing is guided by my intuition that CinemaScope and other widescreen technologies, rather than generating immersive, immediately consumable experiences, are marked by a balance between, on the one hand, emotional engagement and immersion, and, on the other, aesthetic detachment and playful distance. Any further sense of what this alternative consists is flexible and shaped by the films themselves. But given that the works of Negulesco, Edwards, Altman, and Carpenter are inevitably experienced today in viewing conditions far different from those dominant even ten years ago, some further discussion of the rhetoric surrounding widescreen cinema technology is worthwhile prior to a close look at the way aspect ratio and subject matter intersect in the work of these filmmakers.

The word "widescreen" once upon a time referred to something that was, in fact, wider, a scale determined less by artistic practice and more by the historical fact that the adoption of widescreen cinema exhibition, at least in many exhibitors' installations, led to an experience of movies projected on wider screens than had existed before. But mention the word "widescreen" to a casual moviegoer today, and a variety of associations emerge, not many of them to do with the experience of cinema in a theatrical auditorium, or even with a particular range of shapes. Depending on who is uttering the term, the word "widescreen" today might refer to the "wide," 16:9 ratios of flat-screen HDTVs in home theaters, such screens that are now quickly becoming the norm for film viewing. But for younger film viewers, especially, such televisions may not even be thought very "wide"—they are simply there, neither wider nor narrower than any other kind of screen but the type of screen one has grown up with—and perhaps only as adjuncts to the phones and tablets that remain the apparently constant focus of their attention. Another meaning signaled by the contemporary use of the word "widescreen" is its relationship to the use of "letterboxing" to preserve the shape of a film's original format, should it prove wider than the 16:9 ratios of the common television, on home video formats such as DVDs, Blu-ray, and streaming. "Widescreen," further, can today be used in contexts having nothing directly to do with cinema. For example, even the wider monitors of certain computer screens, which in "nearfield" working setups can appear relatively expansive to users who grew up with the square monitors of the first personal computers. The word "widescreen" and its various meanings have in this way become polysemous, its definitions contingent not only on the theatrical experience but also on a variety of home media and personal devices.

Widescreen as a gradually evolving technology, a set of artistic norms governing the use of that technology, and an industry rhetoric selling a certain kind of experience has been the subject of historical study focusing on the evolution of the technology (see again Belton and Cossar). What counts as widescreen as an aesthetic experience, however, has received less commentary in academic work. "Widescreen" in this way marks a range of experiences in relation to a technology perpetually mutating and shifting in terms of the kinds of aesthetic experiences it can enable. I can retrospectively remember an earlier stage in the gradual normalization and domestication of the grand widescreen frame from my childhood in the 1980s, in which wide images would often be presented in small multiplex auditoriums in shopping malls, tiny little rooms that canceled out any sense of widescreen as the transformative marvel it by all accounts was for many in an earlier generation. (As James Spellerberg points out in an important 1985 essay on the perceived ideological effects of CinemaScope technology, the small multiplex auditoriums did not *cause* the assimilation of widescreen in the larger cinematic landscape but were part of a longer process of the technology's domestication; see "CinemaScope and Ideology.") Perhaps because of that reduction in size, my young self did not actually perceive cinematic images as distinctively "widescreen" until the mid-1990s; ironically, it took a small-scale home video format to show me the difference. The fact of diverse aspect ratios became apparent to me when I eventually discovered the difference between pan-and-scan VHS tapes of movies and letterboxed laserdiscs, the latter usually preserving the films' original frames. That laserdisc, a niche home video format, offered the initial lesson to this young viewer about what the most purportedly spectacular of theatrical formats—widescreen—even was, is a trick of history. Perhaps, in a sense, the very form my "discovery" of widescreen took prepared me to see in this technology—even when engaging with widescreen films in a cinema—not an "immersive" format but rather an aesthetic one with a particular shape that was distinctive and visible, thrown into relief against the 4:3 frame of a typical 1990s CRT television, an image marked by borders separating it from the rest of the screen and my immediate surround, not unlike the frame of a painting hanging on a museum wall.

Much of the existing academic discourse on widescreen has focused not on the variously situated aesthetic experiences of wide frames, however, but on the technology's march through film-industrial history. Understood in this way, widescreen is an evolving format, with each "innovative" widescreen picture offering a potential pathbreaking guide to what new possibilities might be discovered in the technology. As the editors of an important collection of essays on widescreen in global cinema write, CinemaScope "was perceived as a process that would augment the cinematic experience, particularly when allied, as it always was early on, with colour and stereophonic sound" (Belton, Hall, and Neale, "Textual Analysis," 59). Anamorphic widescreen, for these authors, is an intensification of the spectacular aspects of cinema that existed in previous decades in narrower aspect ratios and in black and white. Today, the idea of widescreen, when it is employed at all, is often thought of as a means to enable cinema not to evolve but rather to *survive*: the idea of a *big* screen, at least, is still mobilized today by the large cinema chains in North America, which even in the twenty-first century frequently show advertisements preceding the main feature in which viewers are encouraged to remember that the spectacle they are about to consume is only able to really be "experienced" on the vast screen in front of which they are presently seated rather than on the cell phones or tablets that lure users away from the cinema with their convenience and immediacy. But given the ways in which the very word "widescreen" has been thoroughly recontextualized within a range of discourse about media competing with theatrical cinema (television, home video, video games), critical discourse on widescreen is today likely to be determined less by the technology itself (already well established in history and no longer salient in discourse) and more by the critic, who sees the particular use of widescreen as relevant to a discussion about a particular film. While I do not agree with James Spellerberg—writing in 1985 words that very well could have been written in today's fractured and atomized media landscape—that "the contrast of the narrow Academy image to the wide 'Scope image no longer exists" (28), I do believe that only through the viewer's critical and sensual perception of the filmmaker's practice does the ongoing fact of a distinction between the wide image and narrower forms of imagery become salient. While much of the emphasis in the film industry is on "innovations," a fresh exploration of past cinema can nevertheless intensify and sharpen the perceptions and forms of creative thought that fashionable trends and contemporary presentism tend to obscure. As the poet Robert Hass writes, "Perhaps the very loss of context, like the lost context of the animals drawn on the walls of the Lascaux caves, intensifies it" (273).

Francesca Liguoro and Giustina D'Oriano, in their writing on the subject, carry the torch of widescreen's aesthetic power decades after the initial novelty of and industry ballyhoo surrounding widescreen have worn off. They speculate that CinemaScope, and widescreen movies generally, offer us the potential of visual plenitude that, far from overwhelming spectators and sending them into a collective stupor, engages perception aesthetically in a way that remains ongoingly invigorating:

> [The viewer's] eyes can wander around the screen without missing any action, because what s/he is looking at, that is to say the image, is now wandering, too. The free-will guiding the spectator's wandering eye is opposed and complemented by the image's new freedom to wander. This new non-oriented vision gives the spectator license to get lost . . . there is an overflowing of vision. . . . What is shown, i.e., the content of the image, is less important than the fact that it *is* shown. . . . In the new world created by CinemaScope, the spectator is an aware viewer who is looking at an aware surface. (301)

I do not agree with the idea that "what is shown . . . is less important than the fact that it *is* shown"—after all, *Playful Frames* is concerned with the relationship between subject matter and widescreen technique. For me, what *is* shown is a crucial part of the play between aspect ratio and subject matter. Nevertheless, Liguoro and D'Oriano's "wandering" perfectly captures the sort of engaged and relatively free aesthetic experiences that can generate a range of responses to widescreen aesthetics. Wandering, notably, is not quite the same thing as being immersed; we are not locked into the image but are rather free to frolic about it, exploring it, imagining alongside it. Their thoughts also anticipate and prefigure Christian Keathley's idea of widescreen technology as enabling of "panoramic vision," the free-roaming glance of the film lover who desires new revelations and epiphanies in cinema, and whose operating principle is that these new revelations can be found not only in new films but also in an existential engagement with films from the past. Douglas Smith, dialoguing with Keathley's work, has suggested that this "panoramic perception enabled by processes such as widescreen allows the cinephile both to be immersed in the image and to focus freely on the detail within it" (118). Of course, there is no reason to suspect that earlier, narrower Academy ratio cinema could not at times enable "an aware viewer who is looking at an aware surface." But we might still consider CinemaScope, Panavision, and related technologies as soliciting an especially vivid awareness of the potential expanse of a frame, particularly in the hands of filmmakers with a taste for widescreen composition. Liguoro and D'Oriano point to the possibility that, in practice, the work of filmmakers who play in widescreen can generate delightful, provocative aesthetic experiences. And should we seek them out and frame them in our perceptions as distinctive relative to narrower types of cinema—and narrower, more myopic forms of *seeing* cinema—such films can be perceptually and intellectually invigorating, opening up new thoughts about the dizzyingly numerous relationships possible between frame and subject matter.

Playful Frames

I want to offer closing commentary in this introduction on what I broadly conceive a playful wide frame to be. The best evidence of what I mean by playfulness in this book will be found in my discussions of the films themselves, rather than through any grand theory of play articulated here and then grafted onto subsequent discussion. If there is to be a framework for how to think through widescreen, I would prefer to generate it through an encounter with the works of Negulesco, Edwards, Altman, and Carpenter, rather than lugging along an already articulated discursive apparatus to their films. But the idea of the "frame" has a long history in academic film studies—as does the concept of "play" in philosophy—and it is valuable to briefly examine a sample of playful thoughts on the possibilities of the frame.

While I am certainly unable to offer anything approaching a comprehensive overview of philosophies of play in these pages (see Ryall, Russell, and Maclean for more on this subject), I want to briefly gesture toward the idea of play animating this book. Catherine Homan has written perceptively on the activity of the playful spectator in her work on Nietzsche and Gadamer. She begins by pointing to Nietzsche's insistence that "art does not expressly fulfil a function, but instead arises from the superabundance that is beyond strict utility" (100)—a fascinating idea in relation to widescreen, which from its earliest publicity in the industry (especially surrounding the Cinerama format) was considered a medium of spectacle and excess rather than a self-effacing vehicle for classical narrative information, even as the format was eventually domesticated for routine narrative uses. But for Homan, a superabundance of aesthetic pleasure does not necessarily dull or dim the spectator's engagement with the work, as Hollywood's immersive and escapist address to the viewer might. Instead, this superabundance of potential aesthetic pleasure enables, for Homan, a thoughtful engagement with art, for when faced with such excesses the spectator does not simply consume but rather creatively and productively engages with the frame in a contingent encounter (in the case of film, these are various encounters with the ever-shifting and forward-moving play of frames).

Homan's thinking, which repeatedly uses metaphors of space evoking both the content of the film frame—space as creatively imagined fictional world in which we become involved—and space as the four corners of the widescreen frame, the aesthetic canvas on which those worlds are created for viewers, emphasizes the idea that art requires a playful viewer for aesthetic realization to happen:

> We take notice of certain boundaries of the work while playing. We cannot impose any particular meaning on the work, since to do so would be to treat the work as merely instrumental. . . . We take notice of the certain boundaries of the work while playing within the space it opens up for us and, through play, we hold open. In this way, our play with the work opens up a particular world, in the sense that this space is not meaningless, but, like the everyday world, is bound by time and space and is a locus of meaning and understanding. (101)

In widescreen cinema, we "take notice of the certain boundaries of the work" in the form of the aspect ratio and the fact of the frame (for most of the films discussed in these pages, a frame technologically achieved via either CinemaScope or Panavision). And the "certain boundaries" of the work are also at play in the way words are used to describe experiences of widescreen cinema. Take, for example, just a few of the many senses of the word "frame" involved in discussing widescreen cinema: the composition of internal frames within the widescreen frame itself; the narrative "frame" of the story worlds that the filmmakers create within shots and between them; and the "frames" we bestow upon these worlds and film frames creatively, as viewers and critics, as we seek further words to preserve and evoke a certain way of seeing these superabundant arrangements filmmakers have animated for us. Such words about widescreen signal an expansion of our own world rather than a further narrowing of it. Homan's words also offer us the possibility to place immersion and frame-awareness into thoughtful balance, with moments of emotional, and even frame-forgetting, involvement balanced out against relatively detached perceptions of and engagements with the compositional art of the wide image.

I am using "frame" throughout this book, following Homan's sense of the critic's playful perception of frames, much in the way Edward Branigan uses "frame" in his study of the words used to describe cinema, *Projecting a Camera: Language-Games in Film Theory* (2006). Branigan argues that there are many different senses of a "frame" at play in the linguistic gymnastics of film theory: the actual, physically manifest frames that exist in cameras on film shoots; the more abstract aesthetic philosophy implicit in a filmmaker's wielding of a frame; the film frame projected in the cinema; and the metaphoric level on which a "frame" functions as a particular method or "framework" in discussing cinema. Following Branigan's lead in taking "frame" in its various, polyphonic senses, the wide frame in this study might variously be understood as the technological apparatus of widescreen cinema as a materially existing and evolving film technology; the aesthetic and technological frames my four filmmakers employ over the course of their careers; and the various "frames" of film criticism. A range of "widescreen words," deployed at various inflection points throughout this book, are used in ways similar to the usages Branigan points out in relation to "frame": words such as "expansive," "anamorphic," "lateral," "engulfing," and "composition" will refer both to material objects of cinema and to the effects, variously aesthetic and thematic, of widescreen technique. Widescreen is in this way not only a cinematic technique but also a set of film-critical techniques that inform a vocabulary. My preference, however, is to discover these words through encounters with films, situating my play with words in the context of these directors' work with the anamorphic image.

One of the more interesting suggestions of how the widescreen frame can spark an imagination is offered by Charles Affron, whose book *Cinema and Sentiment* remains one of the great works on emotion in cinema studies. In his second chapter, "Thresholds of Feeling," Affron links the film frame—in most of his classical film examples, the Academy aperture (1.37:1) frame used in classical Hollywood cinema—to the larger tradition of the frame in painting. He finds in the frame of both painting and film "comforting evidence that we can bring knowledge into purview, and grasp it with a strength that tests and validates our own powers of perception and imagination. It *holds* the perceiver's attention" (26). Affron notes also a large history of psychoanalytical film criticism on the frame, pointing out that it seems to evince a desire on the viewer's part for mastery. When we witness the frame, and the director's technique of framing, we are also witness to the presence of ourselves, and to our ostensible desire for knowledgeable mastery of the visual world a film presents. The frame, for critics of a Lacanian and Freudian bent, would seem to exist in "the way it is perceived, with the notion that it in fact creates the perceiver" (30). Yet Affron is attentive to the variety of compositional techniques possible within the cinematic image, and in such attentiveness he reminds us how a filmmaker might playfully intervene in our too-confident assurances of our own "mastery" over the film. His book is replete with descriptions of doors, windows, pillars, and other figurations of mise-en-scène that generate framings within the shot that ask their viewer not simply to behold or exercise mastery over the meaning of a frame but to look at other framings, other contingent ways of seeing and living, temporarily hidden, in any given moment within the larger frame. Affron recognizes that even though, in many classical films concerning themselves with domestic, familial dramas, "the frame is the home, the hearth, the context necessary to the fiction's existence and congenial to the human identity at its center" (39), he is also aware of the potential for frames—in particular, widescreen frames—to throw this sense of our comfortable viewing habitat into "meaningful disarrangement" (37). "Even the proportions of *Giant*," Affron writes of the widescreen George Stevens film, "accommodate . . . [a] . . . sense of hearth in the image of a house isolated in the disproportionately wide horizon lines of prairie and CinemaScope. The 'expanse' of feeling is often endlessly variable" (40). It is this expanse of feeling, and of thinking, and a corresponding appreciation of the composition of the frame that situates that expanse in frequently unexpected ways, that compels me to study this quartet of directors in the present book.

1

Jean Negulesco (1900–1993)

CinemaScope Connoisseur

Films discussed in this chapter: *How to Marry a Millionaire* (1953); *Woman's World* (1954); *Three Coins in the Fountain* (1954); *Daddy Long Legs* (1955); *Boy on a Dolphin* (1957); *The Best of Everything* (1959).

Jean Negulesco, born in Romania, was a trained painter, sketch artist, and connoisseur of the fine arts before and during the period in which he became CinemaScope's first great stylist. During the 1940s and early 1950s, Negulesco worked in conventional Academy ratio, in accordance with Hollywood norms. In 1953 and until the end of his career in 1970, he worked exclusively in the widescreen frame and in vivid color (most often in the Technicolor or Deluxe Color systems).[1]

[1] This chapter restricts itself to a discussion of some of Negulesco's widescreen films that are available, at the time of my writing, on DVD or Blu-ray. At present, *The Gift of Love* (1958), *A Certain Smile* (1958), *Jessica* (1962), *The Pleasure Seekers* (1964), *The Invincible Six* (1970), and *Hello-Goodbye* (1970) are not available in good transfers, and I have been unable to screen well-preserved anamorphic 16mm or 35mm prints of these films. I have also excluded *The Rains of Ranchipur* (1955) and *Count Your Blessings* (1959), very interesting films but, for me, lesser achievements than those selected for analysis in this chapter.

His *How to Marry a Millionaire* is the first film shot in CinemaScope, and the second Scope film released to the public, after *The Robe* (1953). *Millionaire* has received copious academic attention as a star text featuring Marilyn Monroe, alongside Lauren Bacall and Betty Grable (see, for example, Konkle; Banner). However, this film and other Negulesco works in widescreen have been not only neglected but derided in scholarly literature. Even David Bordwell, who writes with peerless grace and insight on widescreen, tends to regard *How to Marry a Millionaire* as one of the "innumerable" variety of "Scope items [that] look lumbering and archaic, largely because of the constraints built into the first wave of the technology" (*Poetics of Cinema* 290).

This low regard for Negulesco's cinema dates to the initial releases of the films. Critics writing in the early 1950s, reared on a diet of conventional continuity editing and a narrower aperture, fret about Negulesco's propensity in Scope for shots of long duration, for images and sequences built around the actor's languorous movement across an elongated and lavishly designed frame. For such critics, Negulesco's first Scope films relinquish the supple expressivity of editing. Penelope Houston, on *How to Marry a Millionaire*:

> This type of comedy should be fast, sharp-edged, smoothly paced. In CinemaScope, the effect becomes that of intimate revue played at Drury Lane. Dialogue, booming hollowly from the loudspeakers to left and right of the screen, falls flat; the players prowl aimlessly around rooms the size of tennis courts; subtlety of effect can scarcely be attempted; and a close-up of an aeroplane propeller seems easier to manage than one of the human face. (qtd. in Gibbs 77)

For Houston, the CinemaScope format, with its perceived excess of screen information relative to the conventions of narrower Academy ratio cinema, hobbles screen comedy. But to languish in rather than speed through the cinematic moment is precisely the possibility Negulesco creatively activates in his Scope cinema. As Sam Roggen has shown in his careful study of shot lengths in early widescreen films, the new technology "initially encouraged filmmakers to cut less" (161), a taste for longer shot duration that for Negulesco becomes a permanent trait of his Scope authorship. Negulesco pushed for the use of CinemaScope, in the early 1950s, when it was not clear that any widescreen format would become standard. In an industry handbook of the period, *New Screen Techniques*, Negulesco proselytized for CinemaScope's ability to free the director from thinking "in terms of cuts, dissolves, close-ups and inserts. A director no longer will have to worry about cutting down a scene of enormous scope" and was freed to "visualize . . . scenes in their entirety" (176). It is perhaps unusual for Negulesco to describe his scenes in terms of their "enormous scope," precisely because most of his Scope films eschew grand spectacle for what are, in his hands, leaner forms: melodrama, the musical, and the romantic comedy. Even his touristic films, which indulge in the visual delights of the travelogue—*Three Coins in the Fountain* and *Boy on a Dolphin*—are relatively intimate experiences.

Studio executives initially resisted Negulesco's interest in applying CinemaScope technology to nonspectacular subject matter, but Negulesco saw widescreen in more flexible terms. As the director commented, Scope "makes possible the dramatizing of situations which would be confined and imprisoned on the printed page or in a mouthful of dialogue. No director has the power to portray with montage or long shots the magnitude and spirit of New York City as we have done in this picture [*How to Marry a Millionaire*] with a single shot of the city's skyline at dusk" (qtd. in Negulesco 176).

Negulesco's Aesthetic Gaze

As biographer Michelangelo Capua points out, Negulesco sees in the widescreen format a suitable canvas for the expression of aesthetic cultivation. This sensibility is evident in the presence in his frames of art objects that Negulesco himself collected or created—paintings, sculptures, and sketches. "Like Alfred Hitchcock," Capua writes, "who would 'sign' his films appearing as an extra, Negulesco signed with his artwork. In most of his films, there's at least one of his paintings or sketches included in the set dressing" (2). The inclusion of these objects in Negulesco's films, however, goes beyond set dressing. Negulesco's arrangement of aesthetic objects in the wide frame constitutes a laterally arranged signature inscribing the director's connoisseurship and aesthetic taste within the cultural world his characters navigate. It implies that at least some of these characters share in or aspire toward something resembling Negulesco's taste in and knowledge of the arts.

The inclusion in the mise-en-scène of paintings and other artworks from Negulesco's private collection, which includes works from his own oeuvre as a practicing artist, suggests that Negulesco's approach to CinemaScope is a painterly one. John Belton makes an intriguing contrast between widescreen film, viewed in the cinema, and the frame of traditional painting: "In the history of painting, the canvas painting bordered by a wooden frame replaced the fresco or wall painting during the Renaissance; 1950s cinema reversed this change as the frame gave way to a wall. Though the notion of frame did not (and could not) disappear, it was dramatically redefined" (196). Belton notes that consciousness of the presence of a "frame" (or, in the cinema, a wall or a set of curtains demarcating a projected frame from the auditorium's surround) is always potentially deferred in widescreen spectatorship "because it takes the spectator longer to digest the contents of the frame" (197). Negulesco's cinema encourages the viewer to take a measure of aesthetic distance in relation to onscreen drama even as the films invite pleasure in the beautifully lazy contemplation enabled by this gentle aloofness. For Belton, the new emphasis on duration in Scope cinema means that "the viewer's ability to exhaust the details contained within them tends to be reduced" (197), allowing for precisely this kind of languid, panoramic scanning of the image for visual information, a Scope aestheticism. In Negulesco, full narrative immersion in the Scope frame is also deferred by the presence of internal frames—specifically, canvases and other aesthetic objects—that invite and tickle the eye, works of art that, relatively free of narrative burden, beguile the viewer's attention away from efficient and obedient consumption of narrative information. These aesthetic objects in Negulesco encourage languorous viewing, a detached lavishment.

Negulesco's connoisseurship of the arts, in his Scope movies, is driven by the connoisseur's desire to display collected works, as well as the tourist's globe-trotting around the world to see artworks in situ. The director's aesthetic gaze in his early widescreen films activates a leisurely attention that defers fulfillment of expectation and eschews classical narrative economy. Take, for example, a series of moments involving aesthetic objects in *Three Coins in the Fountain*, Negulesco's second film in CinemaScope, from 1954. The success of *Three Coins* convinced Twentieth Century-Fox that more intimate subjects could live in the wide frame (see Bordwell, *Poetics of Cinema* 287). The cinematographer on *Three Coins in the Fountain* (shot in 2.55:1 aspect ratio) was Hollywood and widescreen veteran Milton Krasner, who would work several times with Negulesco on CinemaScope projects. Their other collaborations include *The Rains of Ranchipur* (1955), *Boy on a Dolphin* (discussed later in this chapter), *A Certain Smile* (1958), and *Count Your Blessings* (1959). The film opens with shots of fountains and parks in and around Rome, as Frank Sinatra sings the title song over the soundtrack. Sinatra's croon ravishes the frame, swooning romance engulfing the image. This is the closest any of Negulesco's Scope films comes to pure travelogue, the wide image romantically soaking in the visual delights of fountains, sculptures, and landscapes. "Each one seeking happiness," Sinatra serenades in his song about lovers, but "each one" is not yet a character in a narrative. The viewer is free to indulge in Negulesco and Krasner's Scope compositions depicting artworks in situ in a way unbeleaguered by story's encumbrance. What most interests Negulesco's gaze in these opening shots are the aesthetic objects encountered in Rome: the Fountain of the Naiads in the Piazza della Repubblica; to the Fontana del Nettuno, on the Piazza del Popolo; to the Villa d'Este, Tivoli (sixteen miles east of Rome); and then the Fontana dell'Ovato and the Neptune Fountain and Water Organ. Each image is bathed in sunlight and shadows, throwing into relief the musculature of the sculpted figures arranged around many of these fountains. Although the sequence unfurls as a montage, bringing us from fountain to fountain and from landscape to landscape, some shots imply the presence of a curious, searching subjectivity. The opening travelogue is nevertheless fully absent the characters, who once introduced will in the main lack the aesthetic curiosity signaled by Negulesco and Krasner's play with Scope framing. The arrival of the characters brings the burdens of narrative, desire, and projection, afflictions lightly dissipated in the Sinatra song accompanying these travelogue shots, a song that flirts with desire without committing to any particular manifestation of it. In rhyme with Sinatra's tune, Negulesco's aesthetic gaze distances us from goal-driven passions, preparing us to discover opportunities to defer narrative fulfillment through the wide image. The parade of Scope compositions in the title sequence, images that function as a showcase for the musculature, implied motion, and patinas of beautiful sculptures, has tutored us to see Negulesco's characters, and their various interpersonal entanglements, as also kinds of aesthetic figures, whose existence as goal-driven beings might be put into counterpoint by eyes hungry not for developments in story but for gradations of figural placement, gesture, composition, and movement.

Maggie McNamara's character in the film, Maria, is nevertheless interesting in relation to all this, in that her goals form a kind of narrative clothesline on which Negulesco can indulge his aesthetic gaze. Maria is a visitor in Rome who becomes warily enamored of Prince Dino di Cessi (Louis Jourdan), an impossibly handsome aristocrat with a taste for modernist painting. With the encouragement of Miss Frances, Maria feigns sophistication in matters of music and painting to impress the Prince, parroting Dino's aesthetic preferences in a bid to win him over. She deploys her considerable charms in a sequence shot in the Museo Nazionale. Prior to Maria's arrival, Louis Jourdan and Clifton Webb are enjoying a friendly philosophical disagreement over painting: Webb's Shadwell prefers sober classical representation, whereas Jourdan's eye is enticed by neo-impressionism. Negulesco's frame is laterally arranged with the works Shadwell admires, Scope technology functioning as a display for the traditional arts, in implicit riposte to the Prince's suggestion to Shadwell that a taste for old art is the functional equivalent to living without indoor plumbing and running water. Negulesco's subtle camera movement, too, implies how widescreen technique can accommodate the presence of diverse aesthetic preferences. The interest of this camera in aesthetics is of course here most eloquently embodied onscreen not by Jourdan but by Webb, who in this film, as in Negulesco's earlier *Woman's World* (1954) and his later *Boy on a Dolphin* (1957), serves as onscreen surrogate for Negulesco's cultivated aesthetic sensibility. As Maria chatters on about a party where she met Shadwell, Webb politely greets her before shifting his attention toward a bronze statute that becomes visible as Webb turns his body and as the camera moves to reveal the work of classical sculpture presently piquing his interest (Figure 1.1). Webb's committed interests in aesthetic matters go beyond Maria's playful but presumably temporary adoption of the Prince's aesthetic tastes and signal how the visuals of this film use mise-en-scène as a means to display art objects beyond any character's particular goals. Webb's Shadwell only desires to sit and look at artwork he deems beautiful.

This notion that the presence of aesthetic objects within the frame of Negulesco's Scope cinema defers and delays narrative fulfillment becomes especially evident when, after Shadwell's departure from the museum, the Prince and Maria visit the "neo-impressionist" wing, decorated with abstract paintings contrapuntal

FIGURE 1.1 *Three Coins in the Fountain* (Twentieth Century-Fox, 1954). Digital frame enlargement.

to Shadwell's taste for classical works. "This seems to have a feeling of space," Maria opines, playing her part as a cultivated art connoisseur, when the Prince asks her what she thinks of a particular work; the artificiality of Maria's slightly exaggerated hand and arm gestures as she tries on this knowledge of contemporary painting is its own abstraction, a theatrical gesture extended toward the borders of the image and away from the interiority of her "authentic" self. "If I may," the Prince says, near the end of this exchange, impressed by Maria's feigned delight in the impressionists, "sit here and absorb also." The moment suggests that Negulesco's wide frames welcome the characters' experimentation with gesture—an experimentation that, like the art objects surrounding them, delays any immediate narrative fulfillment in its acknowledgment of both the borders of the image in which they perform and the borders of the image on which they gaze. Both the painting's and the film's own fungible representations of form extend to the characters themselves, whose performances play out in a Scope frame in response to paintings and sculptures that have themselves been awarded spatial accommodation across the lateral stretch of Negulesco's compositions.

Aesthetic Display

Given Negulesco's background as an artist, the idea of the widescreen frame as an opportunity for aesthetic display is crucial to understanding his particular, aesthetic use of Scope technology. Negulesco was himself a painter of several still life paintings, and this propensity for the carefully situated everyday object inflects his films in ways that go beyond the placement of some of his own paintings in the mise-en-scène. Guy Davenport writes of still life painting that "between the gathering of food and its consumption there is an interval when it is on display" (3), a notion relevant to Negulesco's cinema. Although relatively languorous in their temporal display of art objects, Negulesco's films nevertheless do not typically represent the labor required to gather up objects in and across his wide images. In contradistinction to, say, a widescreen film such as Howard Hawks's *Land of the Pharaohs* (1955), which depicts the enslaved labor toiling to erect the pyramids, Negulesco typically eschews onscreen depiction of the physical strain required to produce the consumer objects his characters desire. It will be notable, for example, that in *Daddy Long Legs* Leslie Caron's character will become a successful ballet dancer, even though the film does not pay much attention to the work required to achieve this. When Negulesco does represent onscreen the expenditure of labor that precedes achievement, as in the characters' arduous search for the titular object in *Boy on a Dolphin*, it is with detachment; physical toil for Negulesco is not a representational subject but a further opportunity for aesthetic display. This idea is emblemized in the delightful sight of Phaedra (Sophia Loren) "working" in her first appearance in *Boy on a Dolphin* as Loren's body double, Scilla Gabel, in a yellow top that matches the yellow flora on the other side of the frame, flutters across and along the lateral stretch and depth of a wide image that engulfs us not in labor but in the pleasures of a big blue expanse. The double Gabel swims in a Hollywood studio tank that itself doubles for the ocean depths off the coast of Greece. Likewise, Negulesco never arrives at that final moment of fulfilled consumption; the capitalistic worlds on display in his films go on and on after the films are over, suggesting that the act of consumption, which the original marketing campaigns of CinemaScope promised to fully satisfy, is never quite complete. His films are located in that delicate interval between labor and fulfillment, work and consumption—offering aesthetic contemplation for those who can afford it. "It is easy to see that still life has been a kind of recreation," Davenport goes on to say, "a jeu d'esprit, for painters. Manet painting a bunch of asparagus is a man on holiday, like Rossini and Mozart having fun writing comic songs, or Picasso doodling on a tablecloth" (9). Negulesco's widescreen films, and their fascination with the luxurious interval, have this sense of a holiday, too; they frequently portray characters in exotic places, looking for fulfillment while taking vacations or going on adventures.

Negulesco will sometimes tuck away his displays, his forms of cinematic still life, in the four equal horizontal quadrants of the rectangular frame (see Deutelbaum on the use of quadrants in 1950s-era CinemaScope). As our eyes take in the visual plenitude of Negulesco's movies, we are actively discovering displays, kinds of still life, within the frame—just as we might spot one of Negulesco's own canvases in the mise-en-scène. Negulesco is, in other words, inviting us to become connoisseurs. In this way, his Scope cinema is marked by an aesthetic gaze, gifting to his spectator a visual space to creatively carve out active aesthetic engagement. This is a detached way of viewing, but one nevertheless intriguingly bound up with the themes of the films. The viewer's own pleasure in Negulesco's aesthetic gaze intersects with the characters' aspirations, entangled as they are with the display of both aesthetic and consumer objects in the narratives. If

Negulesco's aesthetic gaze defers narrative, it still makes palpable the foreshortening of that deferment by the determinative pressures of commodification, which threaten deforming careful consideration of an aesthetic object into immediate "content" consumption. The lure of immersion is still present, and occasionally fulfilled, in Negulesco's widescreen cinema, but the position of the detached aesthete that the films encourage provides, in addition to aesthetic pleasure, a critical counterpoint to immersive experience.

How to Marry a Millionaire

Aesthetic objects—in particular, paintings—are involved with the consumerist desires of the main characters in *How to Marry a Millionaire*. Negulesco's cinematographer on this film was Joseph MacDonald, a Hollywood veteran of films for John Ford, Henry Hathaway, William A. Wellman, and Samuel Fuller, among others, who would go on to work with Negulesco, again in CinemaScope, on *Woman's World* the following year. (The CinemaScope aspect ratio of *Millionaire* is 2.55:1.) Near the end of the film, freshly divorced Mrs. Schatze Page (Lauren Bacall), having decided to marry J. D. Hanley (William Powell)—a man who can provide her with the money and the culture she desires, a man whom she respects but does not love—paces back and forth in a bedroom in a New York City apartment now also the site for her wedding. Behind her, on the walls, laterally arranged across the Scope image, are five framed sketches drawn by Negulesco (his signature can be glimpsed in the corners of some of these sketches; see Figure 1.2). From left to right (and in various states of legibility throughout the scene, given that Negulesco also variously pans his camera left to right and right to left) are abstract sketches of human figures: one, a dancer, ambiguously female, her arms raised in the air, head and long hair cast down, toes balanced on tips; two, a dancer, ambiguously male, with large shoulders and a torso delicately

FIGURE 1.2 *How to Marry a Millionaire* (Twentieth Century-Fox, 1953). Digital frame enlargement.

narrowing into the pelvis, posed as if soaring through the air; three, a shrouded figure, shadowed with and darkened by lines, a mystery, hidden in fabric; and, finally, fourth and fifth, two figures on the right wall opposite Bacall's mirror and desk, two additional shrouded figures, but with hands and arms visible—these could be two figures held in an embrace, cloaked in the abstract intimacy of Negulesco's drawing.

These figures, variously concrete and abstract, beckon our eye to move along the wall, and laterally across the Scope frame, their stilled poses in counterpoint to Bacall's movement in the scene. Brigitte Peucker, in her discussion of the place of painting in films, notes that an actor's pose can parallel the frozen poses of human figures in paintings on walls in the rooms of characters in films: "Since the pose involves a suspension of movement, it suggests an uncanny sense of their lifelessness—even *objecthood*—that links the actors to the world of the dead" (78). Bacall's performance in this sequence, flitting as it does between subjecthood and objecthood, suspension and permanence, stillness and movement, tacitly links her gestures and movements with the various states of the figures captured in Negulesco's sketches. Having decided by this point to marry for money, Bacall's Schatze Page has effectively turned herself into an object—the happy bride, posed in her situation like the figures posed and stilled in the drawings. But she is unhappy and does not particularly want to be a bride on this day despite all the planning that has gone into it. As Negulesco stages and develops movement in his frames, Bacall's predicament becomes variously associated with the sketched figures displayed on the wall.

As the sequence begins, in a long shot, Bacall is framed as a still life: in her white, elegantly frilled wedding dress, leaning against the back of a beige chair, her left hand draped over the edge of the seat, her right leg swinging aimlessly while a maid attends to some stitching on the hem of her dress. Framed this way, it is as if she is the sixth, and most vividly alive, in a series of sketched figures by Negulesco (she is even placed, carefully, in the staging of the shot, between two of these sketches on the wall). Nevertheless, despite the relatively static placement of Bacall in the frame, the fidgety movement of her leg suggests a desire to break out of objecthood, an expression of a subjectivity possessing the unruly desire to move. This concrete desire for movement, this refusal to be permanently stilled, sets Schatze apart from not only the abstract figures on the wall but also the bundles of flowers arranged across the frame: two dozen roses in a vase, on the left side; and, in the foreground, slightly out of focus, pale yellow flowers, waiting to perform their duty as her wedding bouquet, resting on fabric on a brown table. The yellow flowers, placed so close to the lens in these moments, preclude our taking this shot as a mere proscenium, an elongated theatrical stage framed in Scope; this is a carefully designed cinematic frame that holds pictorial dynamism in reserve in anticipation of Schatze breaking out of her poise. And so, the tailoring finished, Bacall gets up (the camera tilting slightly upward, as she walks toward the bouquet in the foreground, to accommodate her height), lamenting

FIGURE 1.3 *How to Marry a Millionaire* (Twentieth Century-Fox, 1953). Digital frame enlargement.

the absence of "those two dingbats" (Grable and Monroe) at her wedding as she moves over to the bouquet of pale yellow flowers, holding them for a moment before placing them back down on the table (Figure 1.3).

These flowers are here transformed by Bacall from still life into expressive object (see Naremore 83–87), helping the actress amplify her character's vacillations. But they are also used in this sequence as a visual conceit, a way for Negulesco to vary his options for movement. The flowers give Bacall a reason to move to the foreground, and a way for Negulesco's composition to balance itself between an emphasis on lateral, left-to-right movement, and movement across the depth of the image, from background to foreground. This strategy is further underscored when Grable arrives to the scene. Here, the staging and framing of the wide image have changed slightly: the table and the yellow bouquet of flowers in the foreground are, for now, not visible, and all planes in the image are now sharply in focus. The cut from the first shot in Bacall's dressing room to this second shot suggests not simply the further development of a narrative but also the intervention of an artist: there is no particular narrative reason for this bouquet of yellow flowers *not* to be here, a point underscored later in the shot when Grable, after entering the room and telling Bacall of her plans to marry the (not wealthy) park ranger she has met, repeats Bacall's movement and walks to the foreground of the frame to reveal (as the camera pulls back slightly to accommodate her movement), once again, the bouquet.

At this, the film cuts back to the foyer, and after a bit of business with William Powell, a bespectacled Marilyn Monroe appears. As James Harvey writes in his description of Monroe's character, Pola, she "is nearly blind without the glasses she refuses, out of vanity, to wear—so she walks into doors, reads books upside down, and so on" (36). But by this point in the movie, she has embraced her spectacles, having met a satisfying lover, a nerdy con man played by David Wayne. And so it is ironic that Pola, once she has finally committed to her corrective lenses, is the only one of the trio not to see or handle the bouquet while in this room: Negulesco frames all the moments with Monroe, including one comical reverse shot during a brief exchange with Bacall, much more tightly than the longer shots that accommodated the foreground placement of the bouquet. The uncertainty over Bacall's impending marriage to Powell already established in the earlier handling of the bouquet, Negulesco can relinquish this visual conceit, for a moment, to let Monroe's stardom command the wide image. Because after all, *How to Marry a Millionaire* is partially a star text, a commodity, in addition to being a carefully designed aesthetic object. When Monroe leaves, a moment later, Bacall returns to the foreground once again, to gather up the bouquet in her hands before deciding—at least momentarily—to go through with the marriage. Yet a sequence ostensibly built around the preparation for legal heterosexual union has been more thoroughly marked by its auteur's interest in a carefully gradated aesthetic arrangement of mise-en-scène and performance in the anamorphic frame.

Other Possibilities

Negulesco's predilection for using the Scope frame as an opportunity for aesthetic display defers narrative and the conventional Hollywood romances often inscribed within them. The stories told in Negulesco's Scope films are light on plot, and even though the narratives, as in so many examples of classical Hollywood cinema, remain concerned with the formation of the heterosexual couple, the construction of this couple in Negulesco's Scope frames becomes notable for the optical plenitude that surrounds it. So while Negulesco's characters are preoccupied with conspicuous consumption and coupling, the director's own taste for the wide image leans toward contemplation and aesthetic detachment.

Thomas Hilgers writes that aesthetically detached works of art "leave us ample space for relating freely to what they present; they allow our capacities to *play* with what they show" (4). Negulesco's compositions invite the contemplation of a range of clashing diegetic personalities and sensibilities rather than serving as invitations to firmly identify with character. Negulesco was an early master of what David Bordwell calls "clothesline staging," in which "shot after shot presents, at various scales, a pair of characters facing each other on the same plane. Bars, lunch counters, and other horizontal settings encourage directors to string several characters across the frame" (*Poetics of Cinema* 307). Negulesco's preference was to move quickly beyond pairs, frequently circling around groups, trios, and indeed entire corporations, situations in which the possibilities of lateral staging are constantly reimagined, recalibrated, and refigured from shot to shot. In such arrangements, he can imagine vivid encounters between human individuals and the faceless corporate companies that, in several of his widescreen films, determine their economic fates and futures. These dramaturgical figurations, particularly in Negulesco's films set in the United States, often play out against and through narrative machinations involving rampant consumerism, phenomena that determine the social structures in which his characters seek success and status. Negulesco views such situations via a patient, painterly gaze, carving out aesthetic spaces of contemplation in public worlds otherwise largely defined by consumerism and immediate pleasure.

Woman's World

Woman's World, like *How to Marry a Millionaire*, marries its fascination with luxurious Scope staging to stories of characters negotiating the worlds of American consumerism. But in an intriguing twist, *Woman's World* centrally involves characters who are not quite right for life in a big city, who finally remain immune to its pricey pleasures. They seek conventional, suburban lives, accomplishments continually deferred and delayed by Negulesco's interest in exploring how the Scope frame can accommodate alternative, urbane pleasures and ambitious achievements. In counterpoint to the earnest, emotional enthusiasm of some of his characters from the American Midwest, Negulesco (working for a second time with *Millionaire* cinematographer Joseph MacDonald, in 2.55:1 aspect ratio) encourages his viewer to take his Scope compositions as an opportunity for various dispassionate, detached perspectives, reminding us that the aesthetic gaze implied in his composition of the Scope frame not only defers narrative and character fulfillment but also quells sloppy or unattractively earnest expressions of emotion.

Negulesco encourages a dispassionate gaze not only through the film's visual design but also through the presence of Clifton Webb, serving once more, as he does in *Three Coins in the Fountain*, as an aloof, aesthete-auteur-surrogate. *Woman's World* begins with a voice-over narration by Webb, playing the character Ernest Gifford, head of Gifford Motors, the corporation around which most of the film's events circle. As Negulesco's camera shifts, in the film's opening sequences, from wide vistas of New York City taken from the bird's-eye-view of a helicopter to a slow downward pan toward the corporate offices, Gifford describes the products his company sells as "luxury on wheels," in the language of the advertising man but without the typically earnest adman's Barnum-and-Bailey mannerisms. Gifford's slightly detached relationship to the products he sells, an attitude of which he is aware—the cars, he says, are "designed to appeal to the snob in everyone"—positions him as a cynical, exhausted aesthete. "And here am I," Webb intones wearily on the soundtrack as he appears on the screen. Webb's detachedness renders his very body, in the Scope frame, not as a site for earnest identification but rather as an object among others to be observed, enhanced by Negulesco's use of process photography in the sequence (a motif of several of the exterior shots throughout the film, and of a few of the interior ones, in a later sequence, in the Gifford factory). This technique serves to abstract the performer slightly from his surround. As Webb gets out of the car, it is evident that the background of New York City, as it is in nearly all the exterior shots featuring the actors in this film, is rear-projected behind him. Rather than distract us, however, this enticing visual abstraction reminds us that the world of *Woman's World* is an artificial one, despite CinemaScope's theoretical ability to get closer and closer to reality, in a Bazinian sense, through its vastness of frame.

Webb's voice-over is only a momentary presence at the beginning of *Woman's World*, as most of the film's narrative events are not seen exactly through his character's point of view; Negulesco's widescreen bounty is too much for any one character, even one as privileged as Gifford, to experience comprehensively. *Woman's World* involves three husbands, each with a wife in tow, visiting the New York City headquarters of Gifford's company in order to jump through the necessary hoops, and to jump through them impressively enough, to convince Gifford to award the man deemed by him most promising a plum corporate job. Gifford is also, it is clear, judging the wives. The display in public spaces performed by the three main women in the film involves their elegant incarnation of social politesse, a display of sartorial taste and carefully poised, socially modulated gesture—the ability to turn themselves into pleasing display, one that will fit right in with the display of capital and luxury the Gifford company is in the business of creating. For the husbands, inhabiting these frames means navigating corporate offices and factories and showrooms with skill and tact, each striving to prove to Gifford that he is the right man for the future of his industry.

Negulesco finds gentle humor in those characters who struggle for a clean fit in this gendered world of urban consumerism and consumption. A recurring motif in *Woman's World* is the abject failure of one of the wives, Katie Baxter (June Allyson), partner to candidate for promotion Bill Baxter (Cornel Wilde), to successfully place herself in Gifford-controlled space in anything but an unfortunate way: Katie's stumbling and bumbling draw attention, but of the wrong kind. The Baxters are from Kansas City, and the film makes painfully evident that Katie knows nothing of how one navigates, physically or socially, a sophisticated milieu. In her hotel room prior to her first public appearance among her husband's coworkers, Katie frets about the purple dress she is wearing, which she fears may not be lavish enough for the occasion. (It is not.) The color of her purple dress, once she and her husband arrive at the party, is thrown into relief against the blues, blacks, and whites of the tuxedos of the men surrounding her, but fails to entice anyone away from the hors d'oeuvres and drinks arranged around a large model of one of the Gifford cars in the background. Katie, the film repeatedly makes clear, is an ideal consumer of corporate products but not possibly one of their skillful creators; neither does she possess the right taste for the kind of luxury Gifford's company is in the business of selling. This becomes apparent later in the film when Katie is entranced by a Macy's store window display of a barbeque set, onto which she projects images of the respective craniums of herself, her husband, and her children. This display is not for the consumer who seeks to become herself part of the elegant or ostentatious display of New

York City life, but rather for the tourist whose object of desire is for practical consumer goods that will return her, eventually, to a comfortable Midwestern abode, a world in which this grill will find its final resting spot. Yet Negulesco's use of one frame superimposed upon another here, internally placing the heads of wholesome June Allyson and her diegetic family as part of the arrangement of pots, tongs, and the outdoor grill, at once binds her closely to consumer objects and serves as a kind of aesthetic display that occupies our attention contradistinctively to Katie's earnest desire. She has become herself a part of a humorous display, but the ostensibly inclusive wide frame has excluded her from getting the joke.

The other wives in the film are more successful at inhabiting these lavish frames, although their motivations for doing so vary. Elizabeth Burns (Lauren Bacall), partner to candidate Sid Burns (Fred MacMurray) knows very well how to navigate the same social circles in which Katie stumbles but has little desire to do so. Sid has worked his way up the corporate ladder at his health's expense, suffering abjectly painful ulcers. When the film begins he has already agreed to divorce Elizabeth because of the ongoing disruption his obsession with corporate advancement poses to the marriage. Elizabeth, however, has also agreed to put on a happy face and perform for one last time the role of pleasant wife while in New York City, so that Sid may have a chance at winning the job. Bacall's inhabitation of Negulesco's frames in *Woman's World* sharply expresses her character's underlying attitude toward this world of consumer objects that has left her marriage in ruins and wrecked her husband's body. A little more than midway through the film, Sid returns to Tony's, a small Italian bistro in Manhattan where he once enjoyed dinner with Elizabeth on a trip to the city early in their relationship. Once seated, Sid sees that Elizabeth is already there. Negulesco cuts to a wide shot of Bacall, smoking a cigarette, framed internally by five colorful paintings of various exotic locales (Firenze, Mantova, Agrigento, Tripoli, and Padova). These internal frames bracketing Bacall are reminiscent of the exotic locales peppering the walls of frustrated James Mason's abode in Nicholas Ray's *Bigger Than Life* (1956), but where in that film posters for tourist spots signaled the suburbanite's repressed desire for expansive exoticism, in *Woman's World* the paintings suggest nostalgia, in this case for an idealized Italy, a fantasy of an exotic past now contained within, rather than lavished across, the wide image. (Tripoli was before 1934 a colony of Italy's in Libya, and by the time of the making of this film would have connoted a lost empire.)

Bacall, unaffected by these connotations even as they suggest something to the viewer about the failure of her marriage, inhabits this image with characteristic cool, while Sid, who joins her for dinner after spotting her at a table across the room, unwisely consumes the same dinner he enjoyed here as a young man, years ago, now exacerbating his ulcers. Negulesco, after these initial wide shots of the characters placing them in the space of the restaurant, shoots and edits the sequence so as to bring us incrementally closer to Elizabeth and Sid, but without ever abstracting them from the surrounding environment of the restaurant that has sparked their nostalgia for an earlier, and more successful, moment in their marriage—and without ever abstracting them from one another, for all the over-the-shoulder shots and reaction shots of Bacall and MacMurray keep the two of them together, a reaction shot of Bacall functioning simultaneously as an oblique profile shot of MacMurray. The sequence is a masterful example of how to establish, in Scope, figures in a larger environment before allowing the actors to explore, in relatively closer shots, the psychological and physical implications of that environment, an environment that remains throughout visually present, in fine-grained conversation.

This moment will be the first of several shots slowly bringing Elizabeth and Sid back together, as Sid eventually agrees to abandon his corporate ambitions: he does not get the job at the end of the film, and he and Elizabeth return home to repair their marriage. In a contrapuntal narrative development, Jerry Talbot (Van Heflin), who ultimately does win the position, will find his marriage to the glamorous Carol Talbot (Arlene Dahl) in tatters. The ability throughout the film of Carol to instantly draw attention to herself in the wide frame becomes a liability to Talbot's candidacy, as she is precisely the sort of social and visual distraction Gifford is looking to avoid as he maintains his company's corporate image. Gifford's gaze is aesthetic in the sense that he is concerned with the sleek look of his cars as consumer products; but as a businessman in charge of a large corporation, he must ultimately value to a greater degree ideals of efficiency and instrumental productivity. Talbot, who would prefer to have his wife not inhabit the social frame of the film with characteristic ostentation and a lavishness that refuses to bend to the will of instrumentalism, eventually tells her as much, bringing an end to their relationship, a move that enables him, albeit unwittingly, to win the Gifford job. One can competitively inhabit Gifford's lavish displays, it soon becomes clear, only after all potential (feminine) disruptions the candidate might make to his corporate image are eliminated.

Negulesco has clearly foreshadowed the film's ending in one of its earlier sequences, in which Gifford takes his three job candidates on a tour of the company headquarters. These spaces of industrial and innovative manufacturing—the most memorable wide displays in *Woman's World*—are populated almost entirely by men. Negulesco's painterly attitude toward these spaces, though, is itself a sensibility that, for the most part, has no correlative in the male behavior at play in *Woman's World*, suggesting that Negulesco is detached from, even as he carefully frames, the male characters in the film. It is appropriate, then, that Gifford's Proving Ground room is filmed with a painterly sense. The Proving Ground is a corporate space in which new innovations in automobile technology are tested, and it is also a room in which Gifford, probing his three candidates as he takes them on a tour of the factory, tasks these men with proving themselves worthy of the job. The Proving Ground is something of a mix between a vast corporate laboratory and an artist's studio, a techno-aesthetic space in

FIGURE 1.4 *Woman's World* (Twentieth Century-Fox, 1954). Digital frame enlargement.

which groups of designers dressed in white coats illustrate drawings of new innovations in car design.

One of the frequent motifs of Negulesco's widescreen cinema, the careful arrangement of beautiful artworks across the breadth and depth of the Scope frame, is beautifully present here, with a dozen illustrations, arranged in an alternating red-and-blue pattern across the walls, of cars drawing our eyes across the upper frame line (Figure 1.4). These illustrations are coupled with carefully arranged models of the same cars—small prototypes arranged on brown pillars beneath the illustrations, and larger ones arranged across the laboratory work space in the middle of the shots. This is an arrangement of the frame that takes an instrumental, corporate space and turns it into a colorful, aesthetic one. Even when one of Gifford's technicians demonstrates, for the three candidates, a new technology in one of these models—the presence of smaller wheels in the middle of the car that allow for easier parallel parking—the purely useful effect of the demonstration is abstracted by Negulesco's focus on the car's sleek, aqua-blue roof and the sloping angles of its lines. If Gifford is underscoring the technological innovations of his corporation for the three men competing for a plum job, Negulesco is turning this space—at least for us, his viewers who can relax with a delight not available to the laborers onscreen—into one of play, of aesthetic delight.

If in this sequence Negulesco is differentiating his languorous aesthete's sensibility from ambitious, goal-minded corporate drive, the director is also using this moment in *Woman's World* to distinguish his own Scope practice from the other major widescreen technology of the 1950s: Cinerama, the curving wide frame that overwhelmed viewers with lavish travelogues and exciting, visceral imagery showcasing the sheer thrill of early widescreen technology. In this sequence in the Proving Ground, Negulesco, however, continues to emphasize the frame of *Woman's World* as a space for the display of instrumental, technological objects transformed into aesthetic bon mots. In the next part of the

FIGURE 1.5 *Woman's World* (Twentieth Century-Fox, 1954). Digital frame enlargement.

sequence, Gifford brings his men into an adjoining room, another lavish display, this time of variously colored (in vivid greens, reds, and blues) car seats arranged across the depth and breadth of the frame, colorful objects that rhyme with the similarly colored curtains draped on the walls behind the characters and that are thrown into particular relief against the flat, monotonous gray of the suits worn by the men positioned laterally across the wide image (Figure 1.5). "Beautiful color combinations," Sid remarks. "I can see why you brought us in here." "Not at all," Gifford corrects him. "I've been looking for a place to sit down." Sid's remark is the only time any of the men in the film make any sort of comment that jibes with Negulesco's own visual interests in *Woman's World*, although it is a notion that Gifford quickly dismisses as he reframes this space in corporate terms (for him, it is a place to rest after a day of work). But what most contradicts Negulesco's own sensibility in this frame—and in every other in this Proving Ground sequence—is not Gifford's comment but rather the occasional cutaways, earlier in the sequence, to shots of an exterior racetrack where Gifford's cars are put through their paces by test drivers. Not only are the unsaturated, relatively naturalistic colors in this exterior footage quite different from the opulent color combinations of the interior Proving Ground sequences, but the placement of the camera in the driver's seat of a car in one of these shots draws our attention to the difference between Negulesco's aesthetic sensibility and the more immersive address employed by filmmakers working in Cinerama. This is precisely the sort of immersive experience Cinerama films sought to achieve, or "the most extreme instances of cinema as pure spectacle, pure sensation, pure experience," as John Belton describes the use of the technology (*Widescreen Cinema* 97). If the Gifford company has tasked itself with creating cars that might offer their own kinds of delirious sensations, Negulesco ultimately distances himself from these desires. In *Woman's World*, Negulesco seeks not to immerse us in the world of the film but rather, with an aesthete's slightly detached eye, to put it on colorful display in the wide image. In doing so he asks us to think about the gendered and normative ways in which corporate display, the kinds of displays *Woman's World* refigures aesthetically, produces and reproduces itself with an efficiency at odds with both aesthetics and the idea of free play itself.

The Best of Everything

Gender in the urban workplace is once again a theme for Negulesco, five years after *Woman's World*, in *The Best of Everything.* Negulesco works on this film with cinematographer William C. Mellor (shooting in 2.35:1 CinemaScope) for the only time; Mellor's other notable widescreen films include *Bad Day at Black Rock* (John Sturges, 1955), *Giant* (George Stevens, 1956), *Love in the Afternoon* (Billy Wilder, 1957), *Peyton Place* (Mark Robson, 1957), and *Compulsion* (Richard Fleischer, 1959). *The Best of Everything* involves a young woman, Caroline Bender (Hope Lange), a freshly hired office secretary at Fabian, a Park Avenue publishing company specializing in mass-market paperback editions of canonical literature. Most of Caroline's labor in this office finds her in the assistance of the tough, seasoned editor Amanda Farrow (Joan Crawford), who keeps coworkers at an icy distance. But Caroline develops close friendships with Gregg Adams (Suzy Parker), a statuesque secretary whose hopes to become an actress are pinned on a troubled love affair with a womanizing theater director, David Savage (Louis Jourdan), and the plucky but naive April Morrison (Diane Baker), who aims to survive in corporate America long enough to land a husband. The office headquarters of Fabian are built on principles of efficiency and hierarchy: three rows of typewriters, all populated by female secretaries, define the space of the middle of the office, while three surrounding walls are dotted by the attractively multicolored doors of the editors of the various publications Fabian produces. Most of these editorial offices, not surprisingly, are populated by men, with the exception of Farrow, who is defined in the social world of the film by her lack of a spouse and a family, essentially making her the female equivalent of the male editor, Fred Shalimar, played by Brian Aherne, who is purportedly married but exudes the randy attitudes of a bachelor.

Although in *The Best of Everything* the female characters find themselves in a hierarchical (and male-dominated) world, there is also the chance to become a part of this world and to potentially redesign it. And the novelty of *The Best of Everything* as a melodrama is that the characters might arrive at this redefinition not through the heightened emotional expressivity conventional to the genre but through ways of moving through richly designed spaces. The film's opening credits sequence signals this idea, laterally, and with a lavender touch, decorating the wide frame with lush cursive text illustrating the names of the actors and crew as, in the background, workers go to their offices in Manhattan. Through the design of this credits sequence, Negulesco is aligning a languorous aesthetic sensibility with female figures, slightly to one side of New York City's bustle.

FIGURE 1.6 *The Best of Everything* (Twentieth Century-Fox, 1959). Digital frame enlargement.

But at certain moments his style will detach itself from character, retreating to increasingly distanced perspectives, his Scope compositions reasserting his position as a sympathetic but aloof aesthete. The mise-en-scène of the office in *The Best of Everything* is sometimes filtered through the sensibility of a particular character, and in such moments Negulesco's sensibility as a visual artist is freely and indirectly mixed with the sensibility and hopes of that character as she climbs the ladder of a workplace, succeeds in professional achievement, and finds romance. At other times, Negulesco's frames in *The Best of Everything* are more objective, looking at the characters like figures in a painting, or correlating their poses and movements to Negulesco's placement of paintings within the frame. Such compositions remind us of the hierarchies, both visual and social, at play in this narrative's world, even as they allow Negulesco a measure of distance from the diegetic power struggles.

Our first glimpse of the Fabian office is both a subjective and an objective one, the cool light of dawn gently touching the blue, orange, taupe, and green-gray surfaces of the office desks aligned across and within the frame (Figure 1.6). In the first moment of this frame, Negulesco presents the empty office through Caroline's perspective, finding in this expansive space a reflection of her own hopes for the new job. The cramped office looks more open now than it ever will again in the film, a space of perceived potential and opportunity rather than predetermined hierarchy. In a moment, Caroline will step into this frame, shifting the image from an interior panorama across which we can imagine her looking with hope and promise, to herself as part of a panorama across which we look, her very presence in the frame now situating her as part of this office's potential but also, at the same time, of its already existing display. She walks along the side of the office, and along the right side of the frame, and glimpses the work spaces lying just beyond the attractively colored doors of Fabian's managing editors; here are other worlds, and other words, spaces in which careers and lives are made, sold, broken. But for now, in Negulesco's visual scheme, they are merely narrow slices of smaller reality within the intertwined expanse of Caroline's vision and Negulesco's own aesthetic sensibility. Caroline's hopes for the future are quietly projected onto this office and onto a corporation whose ways of delimiting and constricting ambition, if we share in Caroline's naivete for a moment, are not yet felt.

Negulesco's particular achievement in the Scope frames of *The Best of Everything* is to show how both subjective and objective identification are bound up with questions of display and power within the wide image: even at moments of subjective identification with a character in Negulesco's Scope imagery, we are still very well aware of her placement as an object, as a display very like one of the paperbacks sold by Fabian, in the film's corporate hierarchy, and in the film's own frame; and at moments of objective identification, when the camera positions us as more detached, Negulesco's frames will witness gradations of emotion that compel us back to momentary subjective alignment and close attention to performance. One such moment occurs when April, the Diane Baker character, encounters editor Fred Shalimar, played by Brian Aherne, in his office. The scene—actually split across two scenes, interrupted briefly with a crosscut to other events—is on the most obvious level a depiction of the unequal gender relations between men and women in this 1950s corporate setting. The frame and the camera movements acknowledge both characters and the charm of Aherne and Baker, the camera following one or the other at certain moments while the frame parses out its attention to each, never overtly declaring its emotional affinities. Negulesco's mobile framing is initially motivated by April's movement into the office and then quickly back toward the door—she nervously believes she has disrupted Shalimar's work. Aherne's editor, it is clear, controls this narrative space, even if control over the aesthetic space of the frame remains with Negulesco. Aherne walks over to the window and opens a curtain—the wide expanse of his luxurious office window rhyming with the wide frame of the film itself—before asking April what might make her a good editor at Fabian. In fact, she has no such ambition, although she has read the work of Shalimar's old friend Eugene O'Neill. Shalimar once cavorted with the likes of O'Neill, he informs April with a touch of exhaustion, for he was himself once a "boy genius." Here the camera, for a moment, begins to move with Aherne as he walks about his office, momentarily fascinated with this man of power—even if his achievements are qualified by the even greater ambition he may have once had for a literary life. The bookshelves laterally arranged across the frame of Shalimar's office suggests a life devoted to books, but to books as commodities; these four bookshelves are full of the products Fabian sells, color-coded paperbacks attracting the eye of the potential buyer. Shalimar attempts to convince April that Fabian, despite its commitment to surface sheen, has a noble philosophy: on these paperback pages she will find printed the great literature through which the press is educating America. For a moment April seems convinced.

But as Aherne and Baker move back over to the window, the rows of bookshelves across the wide frame are replaced by the paneled windows looking out onto the skyscrapers of New York City, represented here in a beautiful matte painting, in the background, in this studio-shot scene. Shalimar's words, like the books we have just seen in the frame, now seem less the product of a devoted literatus and more a backdrop for a businessman making a play for a girl. The camera remains fixed on Aherne in this long take as Baker looks away, chewing on chocolate, enduring the flirtation. She is looking at the great expanse of New York City but is detached, a visual figure in the frame's form rather than a melodramatic avatar of dramatic content. At this, Negulesco cuts to a view from outside the office, the camera now accorded to no human point of view—the frame is apparently floating outside in the New York City sky—with April and Shalimar internally framed within the window's borders. The framing and camera movement earlier in the scene have flitted between April's and Shalimar's points of view, taking each of them alternately as subject and object, as both psychological figure and object of display (in a corporate world, and in a delectably arranged film frame). The window frames them both as display, as constituting just one of the many dramas between men and women presumably taking place in windows very like this one across New York City, the artifice of the display and of the moment underscored by the artificiality of the entire scene (the matte painting that a moment before had illustrated this vista of a New York City of the Hollywood imagination, the internal frame of Aherne and Baker here rhyming with the framed paintings glimpsed a moment before on Shalimar's walls). Although April has become a part of this display, an object like the books, she is unaware of it, speaking of her interest in boys. Shalimar, however, is lecherously aware of it and uses the moment to try to initiate "a bit of fun." She wrestles away from him, her shriek and the film's cut back inside the office—where we are once again aligned with Baker's movement, and where she is once again relatively free of the constrictions of an internal frame—serving to realign us with April as a character. But the alignment is not consistent or exclusive. April's subjectivity, like Shalimar's, is understood, in *The Best of Everything*, in relation to the ways in which she also consents to become an object, another pretty, virgin face decorating a cutthroat company in the business of publishing and marketing pretty, unread books.

If April is unwittingly turned into an object on display here, Suzy Parker's Gregg Adams, an aspiring actress, more consciously and knowingly puts herself on display. Gregg encounters David Savage, played by Louis Jourdan, at the end of a dinner party she arranges for her boss, Amanda Farrow. Weeks earlier, prior to the beginning of the story events presented in the film's plot, Gregg had auditioned for one of David's plays, but he does not remember her. Parker's statuesque height poses a problem of staging for Negulesco: How to prevent her from towering over Hope Lange (five feet two), Diane Baker (five feet six), and indeed even the mighty Joan Crawford (five feet three) in sequences in which Gregg Adams herself has no hierarchal, social power? In the office sequences this is solved

FIGURE 1.7 *The Best of Everything* (Twentieth Century-Fox, 1959). Digital frame enlargement.

through busy staging across the lateral stretch of the image in which all three, and others around them, are kept moving with busywork, walking or variously sitting down or in various stilled moments of contrasting pose; or in shot–reaction shot structures eliding any particular focus on disparate heights between actors (and drawing our attention, at the same time, to the space surrounding them); and through strategies that enable Parker to occasionally sit down or lean over. In the after-party sequence with the shorter Crawford this is mostly solved not only through the sheer force of Crawford's personality but also by Negulesco's strategy of Scope clothesline composition: across Farrow's apartment walls are framed sketches and illustrations of various human figures, their movements and gestures contained by frames in much the same way as any power connoted through Gregg's height is contained by the Crawford character's assertive presence.

With Louis Jourdan, who is six feet tall, Parker's stature is less of a problem. After Gregg meets David at the end of the party, the two of them return to his place. David's apartment is one of the most luxurious in Negulesco's Scope cinema, a habitat of studied, cosmopolitan decadence. Its atmosphere is blue, cold, steely, metallic; laterally arranged along his walls are Asian sculptures, expressionist paintings of abstract human figures and other shapes, an abstract painting of a goblet filled to the brim with knives, and walls of books and scripts (Figure 1.7). Negulesco's presentation of this apartment is cool, angular, and controlled, reflecting David's tendency to throw whatever interior conflicts might torment him onto his work. Gregg is here playing the part of many women before her, the young ingenue staged to play the part of David's lover. (The scene begins with Jourdan on the phone, informing an assistant that an actor has failed to win a part in one of his plays, underscoring the ways in which this apartment is a site for determining theatrical futures.) Yet as in the earlier scene with April and

Shalimar, Negulesco implies Gregg's subjectivity even as David's theatrical apartment renders her an object. Gregg is wearing an orange dress, its warm color and casual sexiness thrown into relief against David's apartment of cold blue, abstract expressivity, and knife edges. Further, although the sequence begins with Gregg positioned, on the left side of the frame, in front of a bedroom—unsubtly signaling the intentions David has in bringing her back to this apartment—it is her movement, and not his, that motivates the mobility of the wide image in this sequence. If these frames constitute David's stage, it is Gregg's way of moving across them, and of admiring them, that most interests Negulesco and motivates his own aesthetic exploration of the scene. And the director's detached aesthetic interest in how Suzy Parker can move across a wide frame is in contradistinction to David's relative disregard for Gregg. He is on the phone in the first part of the sequence, and his eventual movements toward her are not in admiration of her way of walking or moving or gesturing but rather are a means to clinch her, in melodramatic close-up, and in a romance that ultimately leads to her suicide once David abandons her at the end of the film.

In Negulesco's Scope cinema, to fail to be admired, to effectively position oneself as an object in a world of lush exteriority, is to cease to exist. Caroline, of the three women in *The Best of Everything*, finds the most successful balance between objecthood and subjecthood, between the desire to turn oneself into a display but to not do so at the cost of one's being. At the end of the film, she has obtained a job at Fabian as an editor, perched at a similar level in the corporate hierarchy as Fred Shalimar and Amanda Farrow. At the beginning of the final sequence, as Caroline reads a letter to a secretary, Negulesco's framing underscores the length and depth of Caroline's office and Hope Lange's assured way of walking across it. But where earlier sequences in the offices of Shalimar and Farrow had begun with these powerful editors reaffirming their place in a hierarchy, Caroline, although comfortably placed in that regime, occupies it in a more flexibly horizontal manner. (She is also more lenient with her assistant than Farrow is with her earlier in the film.) As Caroline walks out of her office, across the room full of typewriters, and toward the door leading to the outer world, Hope Lange inversely performs her movement from the film's opening sequence, now in possession of a powerful position rather than just potential. Her underlings at their typewriters look up at her with deference, indicating her accomplishments within this firm. But hers is also a somewhat different achievement than Amanda Farrow's (to whom Caroline says goodbye as she leaves in this sequence), since the ending of the movie promises some hope for a personal life that exists outside the limits of the corporation.

After pausing melancholically, for a moment, near the glass doors of Fabian—where she is now framed by Negulesco's camera, looking out at her from inside the office, her existence defined by and framed through Fabian's visually permeable but nevertheless rigid corporate barriers—she heads out to the city's sidewalk. Now the camera, having performed carefully rehearsed pirouettes in the tightly staged and framed interior environs of the Fabian office, loosens itself up in a handheld shot rare in Negulesco's widescreen cinema. The camera finds itself in the open-air contingency of New York City, as dozens of extras, playing city dwellers, flock and run around Hope Lange. The determinations of dramaturgy are soon reimposed on these images, however, as Stephen Boyd's Mike Rice, the junior editor who earlier in the film had caught her eye, runs to meet her. This reassertion and resolution of narrative is matched by Negulesco's resumption of a more stable camera movement in the final image, a crane shot that gathers up Lange and Rice as they walk into the distance. Nevertheless, the contingency of this open-air photography in the city has thrown itself into relief against Negulesco's controlled CinemaScope aesthetic in other scenes. The wide frame opens itself up here to a world beyond the hierarchy of Fabian.

Boy on a Dolphin

Boy on a Dolphin marks Clifton Webb's final appearance in a Jean Negulesco film. His inhabitation of Negulesco's CinemaScope frame once again obliquely functions as an aloof auteur-surrogate. He plays Victor Parmalee, an aesthete and art collector who desires to obtain the film's titular relic, discovered in the opening sequence by a Greek woman named Phaedra, played by Sophia Loren. The cosmopolitan Parmalee competes with the stolidly American Dr. James Calder (Alan Ladd), an archaeologist who yearns to preserve and publicly display historically important artifacts. Negulesco shot *Boy on a Dolphin* in 2.35:1 CinemaScope with cinematographer Milton Krasner, an earlier Negulesco collaborator on *Three Coins in the Fountain*, which like *Boy on a Dolphin* has a sun-kissed, touristic quality. *Boy on a Dolphin*, more than any other Negulesco film, is concerned with laborious, physical work: the task of tracking down this relic from the ocean floor, an effort requiring capital, skill, remarkable expenditure of the body, and cartographic knowledge. Characteristically, Negulesco, in his implied sensibility as auteur of the wide frame, remains aloof from this onscreen physicality, channeling all this labor into visual delectation, once again aligning himself with Webb's own aesthetic detachment.

Rather than guide us through a travelogue of sculpture and landscapes, as the opening sequence of *Three Coins in a Fountain* does, *Boy on a Dolphin*, complemented by Julie London's serenade of the title song, begins cartographically, with a vivid, bold, colorful map of the Greek islands—Hydra, Poros, Delos, Mykonos, and Rhodes, all playing a role in the ensuing narrative. The map introduces the main tension lying behind the film's widescreen compositions, between the image as a vehicle for narrative information and the widescreen image as source of visual pleasure. The map has an obvious instrumental value: it shows us where the film's events will exotically occur. But it is also aesthetic: it is replete with decorative, beautifully illustrated starfish, ships, and dolphins, the frame enticing us with pleasures beyond story. This map—and its colors, style, and playful figures—resembles a child's storybook writ large across the screen. Additional bold and colorful title cards precede journeys in this sequence to each of the islands, one after the other—first Rhodes, the main setting of the film (where we will shortly meet Phaedra), dotted by lighthouses, windmills, and fishing ships; Delos, alive with historical ruins and artifacts, some of which document the legendary dolphin of the film's title; Mykonos, with more windmills, and with Christian crosses decorating several abodes; Poros, the smallest of these islands, dotted with shops and restaurants and docked fishing boats; and Hydra, represented not by land but by water, as the camera takes us to where no tourist might easily go: the ocean floor, where rests the boy on a dolphin, a fabled sculptural ruin from the mast of a sunken ship, the object desired by all the film's characters.

Phaedra is first seen, after the opening titles, gliding in water across the Scope frame and gathering up artifacts from the ocean floor as colorful fish swim by her in the foreground (in most of these shots she is played not by Loren but by Loren's underwater body double, Scilla Gabel). She spies down here the titular artifact but cannot free it from the wreckage, and so the effort to acquire this precious object of her heritage becomes her obsession. This vision of Phaedra figured against the artifice of a Hollywood ocean complements the colorful maps from the opening sequence. The character is lavished in a bright yellow top thrown into relief against the dark blues, browns, and burnt oranges of the ocean floor, an impressive commitment to underwater fashion rhyming with the vivid colors of the fish swimming about her. Although Phaedra here discovers her narrative goal—one that, like the goals of other women in Negulesco's cinema, will involve complex relationships with powerful men—in some sense the purpose of the film will be to restore this ocean relic to the status that it occupies in these opening images: as part of a colorful, painterly aesthetic display rather than as a figure that carries significant narrative meaning.

Calder, when Phaedra later meets him, is assessing a small museum of ancient objects at the Acropolis in Athens, explaining to an associate why this humble gallery cannot possibly compete with lavish museums in Paris and Rome. The ruins of ancient Greece, he explains, have been depleted by thieves and charlatans, robbing this gallery of erstwhile delights. Calder is a figure of assumed authenticity, a narrative correlative for the film's insistence on authentic environments (and a counterpoint to its parallel fascination with aesthetics and artificiality). Phaedra's initial conversations with Calder, as she tries to convince him to help her retrieve the boy on the dolphin, are figured among the ruins of Athens, the display of Loren and Ladd as a romantic couple becoming part of a lavishly appointed continuum of artistic creation. Her first encounter with Parmalee, by contrast, occurs amid the hustle and bustle of a modern restaurant, near the Gardens of Zappeion in Athens. The slope and angles of its modernist architecture are a contrast to the fragment of rock on which Phaedra sits, contemplating a pair of new shoes that she has bought for her meeting with the connoisseur. Parmalee, when she encounters him, is enjoying a cocktail alone at a table. Phaedra joins him, disrupting his solitude; he is about to leave, but her mentioning of Calder piques his interest. Webb brings a drink to his lip, his presence in this restaurant figured against the horizon line of the ocean in the frame, his cosmopolitan sophistication in light counterpoint to the expanse of ocean in which the valued boy of the film's title lies submerged.

Given Negulesco's implicit identification, as a cinematic aesthete, with Webb's presence in the widescreen frame, *Boy on a Dolphin* is not satisfied with a judgment of Calder as a figure of virtuous goodness and Parmalee as selfish cad. Calder's assumption of values of authenticity and rightful ownership are qualified by the high artifice of Ladd's cinematic figuration against the backdrop of Athens alongside Loren; both are products of the international star system that shaped the making of many CinemaScope films during this period. Calder's assumption of virtue is also qualified by Parmalee's reminder to us that Calder is a military figure defined not by personal taste but by devotion to nation and power, whose own declaration of moral goodness claims the wide image as a space for colonialist expansion and possession. Webb, meanwhile, embodies values of connoisseurship that, as in his other appearances for Negulesco, takes the wide image as a space in which he may make aesthetic judgments and strike distinctive, detached poses of knowing sophistication informed not by nation but by individually crafted cultivation.

A sequence roughly a third of the way through the movie provides an example of the different ways in which Webb and Ladd incarnate Negulesco's Scope imagery. The two are seen together in a clothesline image in a monastery library to which Parmalee has journeyed to research important information about the boy on a dolphin. In contrast to Webb's casual, relatively aimless stroll through Rome near the end of *Three Coins in a Fountain*, his journey across the beautiful and mountainous landscapes of Greece occurs via automobile and is driven by a specific goal: he is on a quest for knowledge of the lost relic. But appropriately for Webb's aloof persona, *Boy on a Dolphin* does not immerse us in the physical nature of his journey but rather displays from a distance the blue and gray vistas of sky and mountain, in ways that free Parmalee himself of any strenuous effort. Although he travels across rough and rocky terrain, Parmalee remains unruffled; Webb inhabits these beautiful images of landscape precisely *as image*, his remote presence and detached sensibility bestowing on these widescreen vistas of the Greek countryside a painterly, aesthetic value. These landscapes are here for him, and for us via him (seated, as we are, and as he is, to view), to consider with cultivated intelligence; their expanse is not reduced to narrative or laborious import but via Webb's presence become frames of pleasure. *Boy on a Dolphin* indeed spends so much time lingering in and around these beautiful frames that it defuses any expectation of a melodramatic encounter with Alan Ladd once Webb has finally reached the monastery library.

Once Parmalee arrives, he finds Calder, already busy at work, researching, having found the information Parmalee seeks. The provisional goal of Webb's journey through the images already achieved, the mise-en-scène of this ornate library, with Ladd and Webb arranged laterally near the foreground, becomes its own attraction. This frame balances books as narrative content—the information Calder has gleaned will help the men find the relic—and books as aesthetic objects, enjoyed for their visible textures. Calder as a vehicle for narrative power, and Webb as advocate for the delectable image, become surrogates for the film's own tension between beautiful Scope imagery and narrative accumulation. In this scene, Webb's sensibility dominates, as this library's vast array of dusty volumes serves primarily a visual and aesthetic purpose rather than a narrative one. While Calder remains diligently focused on work, Parmalee looks at books in a rather detached way, avoiding the hard work, the labor, of selecting the important documents—Calder, before Parmalee's arrival, has already done this—and instead appreciates the fact that his rival's work has relieved him of any need to get his fingers dusty. The lateral staging draws a superficial equivalency between the two characters, but Webb's performance detaches Parmalee from Calder's earnest expenditure of effort. The moment becomes less about labor, our main interest now in the lateral arrangement of two distinctly different male personalities across a wide frame. Like Parmalee, we viewers, having come to *Boy on a Dolphin* for our own pleasure and in our own leisure time, are like Webb not to get our fingers dusty here.

Boy on a Dolphin proceeds toward a narrative culmination in which Alan Ladd and Sophia Loren form a romantic couple, and in which the boy on a dolphin, in the penultimate sequence pried from Webb's acquisitive hands, is restored as an icon of Greek heritage. The final clinch between Ladd and Loren is among the most artificial and forced in Negulesco's Scope cinema: as Loren walks away from Ladd, the film uses an increased frame rate—a tonally jarring effect, like something out of a slapstick silent comedy—to rush through their moment of embrace. Here the panorama of the Scope imagery is foreclosed in favor of a rushed coupling that literally compresses our ability to scan the frame via a hurried frame rate. This strangely artificial star coupling, something Webb's Parmalee would himself probably scoff at, collapses the expansive potential of the Scope frame itself onto the predetermined expectations of the romance genre and is very much at odds with the general tendency toward deferment and delay otherwise prioritized throughout much of Negulesco's widescreen cinema. But its very oddness makes it an exception in Negulesco's widescreen oeuvre. Negulesco's autobiography reveals that this "natural" cinematic pairing of Ladd with Loren is in fact a highly stylized construction; Loren was four inches taller than Ladd, and trenches had to be dug near where she would stand to accommodate Ladd's lack of stature. Negulesco even recalls that for certain behind-the-shoulder conversation sequences, he used a taller and more muscular body double to mimic the movement of a talking head while Ladd, sitting next to the camera, spoke his lines offscreen (251).

Although Negulesco does not mention it in his memoir, it is likely that similarly illusive staging tactics were used during the film's few conversation scenes shared between Ladd and Clifton Webb, who towers a full five inches above Ladd. But unlike the rushed frame rate that forces the Loren-Ladd clinch at the end of the movie, the most memorable moments shared between Webb and Ladd are allowed to unfold in relatively natural duration. In one such frame, Webb is seen by the ocean painting a picture of architecture built along the shoreline. In the staging of the shot Webb's gaze is positioned not toward these homes but toward Ladd, as if it were the actor, rather than architecture or the boy on the dolphin, that Webb's Parmalee most desired to gaze upon and to paint. That Ladd's onscreen virility was in part the product of cinematic illusion is every bit as much in play here in the scenes with Webb as it is in the scenes with Loren. This frame, in which Ladd is arranged as a figure of desire in the widescreen image and Webb is positioned as his eager viewer, reminds us of the nonnormative desires and relative detachment Clifton Webb's presence repeatedly signals in Negulesco's Scope cinema. If Negulesco's CinemaScope oeuvre is one marked throughout by the deferral of narrative consumption, Webb's presence opens up its own alternative space within the flow of that languid duration. Webb is the ideal diegetic spectator of Negulesco's panorama, selecting from what lies before him the delights that most pique his gaze—in this case, Alan Ladd. This is a glimpse into another way of seeing and living not frequently incarnate in Hollywood films of this period, one enabled by Negulesco's delectable fashioning of the wide image.

Daddy Long Legs

Daddy Long Legs, Negulesco's only Scope musical, was shot by cinematographer Leon Shamroy, a veteran of several lavishly appointed productions, including *The Robe* (Henry Koster, 1953), *The King and I* (Walter Lang, 1956), *The Girl Can't Help It* (Frank Tashlin, 1956), *Desk Set* (Walter Lang, 1957), *Porgy and Bess* (shot in Todd-AO for Otto Preminger in 1959), *Beloved Infidel* (Henry King, 1959), and *Cleopatra* (shot in Todd-AO for Joseph L. Mankiewicz in 1963), among others. The film tells the story of Jervis Pendleton III, played by Fred Astaire, an extravagantly wealthy American who one day, when his car breaks down in the French countryside, espies from afar a spirited French orphan, Julie Andre, played by Leslie Caron. Charmed by her talent, beauty, and poise, and with money to burn, Pendleton schemes to become Julie's secret benefactor. His money will enable her to go to school in the United States. Eventually the two will meet again, at a high society ball, and fall in love, despite Julie's ongoing lack of knowledge in regard to Pendleton's identity as her benefactor; Jervis initially envisions himself only as a father or a patriarch and must do quite a bit of work to inhabit the frame as a lover and an equal.

Even though Astaire is not as prickly or subversive a presence as Clifton Webb, he is nevertheless here cast as another of Negulesco's aesthete surrogates, a character who gradually learns how to commit himself to another person and how to live meaningfully in public even though nearly every grain of his personality bends toward solitude. *Daddy Long Legs* is in this way a widescreen film about learning how to go outside, a Scope interrogation into the various possibilities of moving, with careful deliberation, from a private position to a public inhabitation of the wider, surrounding image through dance, song, and merriment. Before we meet him, Jervis is cloistered in a lavish room tucked away in a private corner, not open to the public, of his art gallery showcasing to the world various heirlooms and precious aesthetic objects collected by generations of Pendleton patriarchs. Even this public gesture is cloaked in privacy, though: the gallery is "open to the public," a sign at the beginning of the film informs us, but "by invitation only." "This is a Renoir," we overhear a tour guide say to a group of gallery visitors as Jervis's assistant, Griggs (Fred Clark), walks across the frame to meet his boss. The frame depicting the Pendleton Art Gallery is lavishly decorated with paintings and statues and other precious objects. "Renoir is famous for his paintings of children," the guide informs the group, and us, as Negulesco cuts to a shot showing the painting the guide is describing, "but this is one of his finest." The painting in question is Renoir's portrait (ca. 1878) of the French actress Léontine Pauline Jeanne Samary, a star of the Comédie-Française. One set of gallery notes not terribly different in tone from Negulesco's tour guide emphasizes how the "triangular arrangement of her torso and the fluid contours of her limbs reinforce the serenity of the pose without the distracting theatrical trappings" ("Notes on *Mlle. Jeanne Samary*"). The portrait abstracts the performer from her environment, her body and pose imagined apart, in her own private world, from the surrounding mise-en-scène. As the tour guide describes a Corot painting of a natural landscape—"In this painting the foliage, far more than any signature, proclaims the painter. Note the rich use of browns and yellows"—Griggs opens the door to Jervis's private office, jazz yelping within. Unruly trumpet and drums, saying damnation to distance and inviting immersive auditory experience, interfere with the guide's dispassionate description of the rich color in the painting. The flow of sound from the door suggests the presence of porous boundaries in this gallery, in counterpoint to the calm and abstract serenity figured in the paintings. The sheer energy of sound and vibration disrupt, for a moment, a sense of the Scope frame as an occasion for aesthetic detachment. Any of the Clifton Webb characters we met earlier in this chapter would be horrified.

The moment obliquely aligns Jervis (still unseen up to this point) with a desire to break free of calm contemplation and into the energy of modern life. Astaire is therefore positioned to shake Negulesco's cinema out of its tendency toward Webb-like aesthetic detachment, even as the character Astaire plays is initially a bit reserved. Negulesco in turn reminds us that energy and immersion can themselves become a subject for painterly, detached appreciation. So no cut brings us to Jervis's private office just yet, the film holding the jazzy exuberance of the music at a slight remove. Instead, the frame follows the tour guide as he leads

FIGURE 1.8 *Daddy Long Legs* (Twentieth Century-Fox, 1955). Digital frame enlargement.

the group through the Pendleton Art Gallery, observing a series of paintings of Pendleton patriarchs arranged across the far wall. The first two are painted in the classical style; and while the grayed and angular Jervis Pendleton I bears little resemblance to Astaire, the second portrait, of Pendleton II, more closely resembles the star, and in a traditional representational form (Figure 1.8). The guide's description of these two paintings is interrupted by an unruly member of the tour group, a matronly woman who is sneaking up the stairs to peek into the room that seems to have been the source of the jazz. The guide beckons her to return to her proper place in the group; despite the music's implicit invitation, she is not allowed inside the private offices of Jervis Pendleton III. But the guide can introduce her, the rest of the group, and indeed us to Pendleton III through his modernist likeness in the third of the portraits, toward which the camera now pans, a Fred Astaire done in abstractly expressionist colors, angular edges, and the jagged shapes of a Picasso. "As you can see, he has broken with the family tradition somewhat," the guide observes.

Unlike the matron who cannot enter the room, Negulesco's viewer has the privilege to travel there in a match cut, from this image of the abstract rendering of Astaire to a shot of Astaire in his office. Jervis is holed up here, smoking a pipe, playing the drums in accompaniment to the jazz trumpet on a record. But despite the energy of the jazz, a counterpoint to the staid journey through an art gallery in the preceding sequence, Jervis is not altogether "modern." Internally framing him on the left and right side of the frame are ancient sculptures—a horse's head, and an elongated human figure—reminding us of his inheritance. And Astaire himself evokes, at the time of this film's making, two prior decades of his dancing work in classical, Academy ratio cinema. Yet Negulesco's modern Scope frames will soon adapt themselves to the energy and brash talent already vibrating in the scene as Astaire sits drumming. Griggs, who has entered the room behind Astaire, walks across to answer the phone, affording us the opportunity to see that this room is mostly empty, dotted here and there with framed canvases but mostly assuming the appearance of a well-appointed corporate boardroom. But this emptiness soon becomes replete and full as the room reveals its true purpose as a makeshift stage across which Astaire will dance: in a moment, to the disgruntlement of the business-oriented Griggs, Astaire will treat this office as a dance floor, tapping his way across the floor and the table, now dancing rather than drumming to the recording of the trumpet. The scene is a demonstration of Negulesco's adaptation of his Scope compositions to the talent of the performer: the wide frame becomes a proscenium for Astaire's dancing and tapping, his feet always in view. Astaire enjoys the full length of the frame, moving from left to right, from a table in the middle of the room back to the drum kit, in turn adapting his way of moving to the lateral space of a Scope composition. The scene also prefigures, in condensed form, the trajectory of Astaire's character as a whole: from a man who remains in a private and cloistered position of aesthetic enjoyment to one who expands to enjoy the entire frame around him. But he will need a partner to draw him out to that expanse.

Leslie Caron's first dance steps in this film are more hesitant than Astaire's, as befits the role she plays as a beneficiary and a pupil. Her Julie Andre has natural talent, evident in the staging of a nighttime shot in the French countryside, shortly after Julie learns she is the beneficiary of Jervis's financial generosity. With the younger orphans asleep, Caron enjoys a moment of solitude in which she wonders about who her "Daddy Long Legs" (a pet nickname given to her benefactor) might be. The composition of the frame renders vibrant a mix of natural and highly artificial, painterly elements. On the left, a pile of wood in wait for stove and furnace; a blue-gray wall choked by ivy; on the right, a matte painting behind a table and chairs, depicting other rooms in the orphanage; and in the middle, Caron, standing in front of a chalkboard, on which is scribbled the day's lessons. The frame is awash in blue—and Caron wears a blue shirt, in rhyme with the evening light. As the camera tracks closer to her and the chalkboard, she draws an image of her imagined "Daddy Long Legs," in the form of a comic stick figure in a top hat. As the music swells on the soundtrack, Caron turns around, a more serious expression passing across her face. This gesture toward maturity will find eventual fulfillment through dance and music, through the way Caron moves across the wide frame. A chorus of female singers appears on the soundtrack, singing of Julie's benefactor. From here, Caron's performance authorizes Negulesco's camera movements, following her as she fulfills her nightly duties, such as extinguishing the exterior lights of the orphanage. But Caron also gestures and moves in ways expressive of a personality that will soon be free of such labor: jumping through a swing that rocks in the middle of the frame; teetering on the edge of a bench near a table, as if on the edge of a cliff; and lovingly decorating the harness of a cow—the happy bovine revealed, on the edge of the frame, by Caron's movement there—with a red flower she has plucked from a nearby pot. Caron's command of the wide image in this sequence rhymes with

FIGURE 1.9 *Daddy Long Legs* (Twentieth Century-Fox, 1955). Digital frame enlargement.

Astaire's own in the earlier office space, although Julie's movements are more hesitant than Jervis's bold, brash, jazzy jumps. The film's frames have generated a creative and performative problem: How to bring these two very different types together, in a wide space that can readily accommodate their differences and their separation, but which also promises new possibilities that could see them joined together?

Julie proceeds to America, where she is educated pro bono at Jervis's alma mater. In this way she becomes part of the upper crust of social life. Julie continues to pine for her benefactor, though. She writes letters to him, but Jervis does not read them and does not respond. They are collected rather forlornly by Jervis's secretary, Alicia Pritchard, played by Thelma Ritter with reliable poignancy. While Julie engages in society, Jervis remains relatively cloistered in his art gallery office. These sequences contain an interesting and prescient figuration by Negulesco of the domestication, and instrumentalization, of the widescreen frame for purposes other than art. Within a center wall in Astaire's lavish office is installed a widescreen monitor, an object that is not at all unlike, in its size and shape, the various widescreen televisions many viewers of the twenty-first century, even those not privileged with the wealth of a Jervis, possess in their own homes (Figure 1.9). In 1955, to have a lavish Scope frame in one's private work space, however, even when one was a character in a Hollywood movie, was extravagant novelty. On this screen Pendleton busies himself with viewing designs of various Pendleton marketing campaigns (for an airline he happens to own); these colorful illustrations are not unlike what we might see in the corporate worlds of *Woman's World* or *The Best of Everything*. Griggs interrupts Jervis's viewing of these slides: "A corporation is not a person!" he proclaims, reminding Jervis of his personal obligation to the now not-so-young orphan, Julie, who is benefiting from his financial largesse but yearns for human connection. Astaire sits here in Negulesco's frame with his feet up on a table in front of the monitor, his pose of leisure reminding us of his flippant attitude toward his corporate duties. "I'll vote for that! Where do I register?" Astaire says to Griggs, his feet here (the blue matching the blue of the earlier scene with Caron, a detail in the frame already quietly linking the two of them) also signaling that Negulesco's wide frame, a frame of cinema and of art, is something other than the domesticated, private screen on which Jervis looks at these marketing materials. The ads sell a product, but they are not a space across which one might dance or freely dream. Astaire's feet remind us of the aesthetic purpose of Negulesco's frames, the proscenium they will form for Astaire's and Caron's talents, talents that in turn prompt us to dream along and away.

Dream sequences are crucial to the film, two such sequences conveying important facets of both Jervis's and Julie's hopes and fears and expressing something of the painterly and the magical that the widescreen monitor in Jervis's office, displaying only consumerist advertisements, can only vulgarize. They also work to open up a set of possibilities, not all of which are taken up by the main narrative line, and in this they serve as exemplary instances of Negulesco's preference for the temporary deferment of narrative fulfillment and progression in his Scope cinema. Thelma Ritter delivers Julie's pile of unanswered letters to Jervis's office, and after she leaves he begins privately to read them. The letters contain a series of descriptions from Julie imagining who her benefactor might be, what this American surrogate father might look like, how he might move and dance. We hear not very much of the literal contents of Julie's letters—Caron's voice-over serves only as a brief transitional device, taking us to different variations of her dreamed daddy. Instead, we are treated to a series of Astaire numbers that project who Caron imagines her "Daddy Long Legs" to be. In the first sequence, Astaire, dressed as a baroque cowboy in white hat, red kerchief, elegant black jacket and white shirt, with white-striped gray pants and glossy yellow cowboy boots, takes on the comical appearance of an outrageously stereotypical American man a young French woman might conjure in fantasy. Behind Astaire on this yellow dance floor looms a giant, one-million-dollar bill, painted on the rear wall of the set, as Astaire mimes the words to a Texan variation of the film's title song. Caron's words, however, interrupt the honky-tonk, taking us to the next stretch of dream, with Astaire now imagined as an international playboy, cowboy hat and boots discarded for the sartorial sensibility of an aristocrat. Astaire walks toward the camera as it tracks back, as swooning women, emerging from both edges of the frame, faint in his presence. "I don't like him like this!" Caron suddenly interrupts; and a new orchestration of Astaire is provided, now a "guardian angel" whose movements through the wide frame are more gentlemanly and guarded. This is closer to the Astaire we expect. Here, in this dream, Caron joins him, on a set with small corner shops vividly decorated in pastel pinks, yellows, blues, and greens. Astaire's movement takes us from a long shot of this set to an internal frame, inside an ice cream shop, as Caron pines for a cone. Astaire plucks an ice cream from the shop, and Caron follows his luring of her to the treat, her eyes fixed on the delight the guardian angel is offering but not quite yet glimpsing the angel himself. Astaire soon takes her out of her isolated internal frame, and as Negulesco cuts to a long shot, they dance across the pastel-colored set, their two fantasies of one another merged, for the first time in the film, in a dance across the wide image.

This sequence's mix of Julie's words and the frame's visual manifestation of the imagined Jervis her words project precede the actual meeting of the two characters. But this is not the film's last exploration of subjectivity and fantasy in the Scope frame. Near the end of *Daddy Long Legs*, Jervis breaks off his budding romance with Julie, who is still not aware Pendleton is her benefactor. Via a series of narrative complications, she has taken him to be the uncle of an officiary at her university. Jervis has been encouraged by the American ambassador to France, Williamson (Larry Keating), to break off his relationship with the young French girl; the romance, for Williamson, takes on the appearance of an abuse of his office (Williamson has helped Jervis arrange Julie's education and financial support in America). After the relationship ends, Julie continues to read of Jervis's travels across the globe in the newspaper. She collects clippings, arranging them across the floor, mournfully gazing at them. She begins to write another missive to her "Daddy Long Legs," asking him if she might see him in person for advice in the wake of her broken engagement. While writing this letter, however, Julie falls asleep, and the film transitions to a twelve-minute ballet, a solo number by Caron, punctuated only occasionally by a detached Astaire, an audience member and not her costar in this fantasy.

In the first number, Julie is fashioned as a globe-trotting dancer, traveling to Paris, then Hong Kong, then Rio de Janeiro, all in search of Jervis. In a kind of inversion of the earlier dream sequence in which Julie imagined her "Daddy Long Legs" as three different types of men, in this sequence Julie dreams herself as a trio of different women. In Paris, she is a professional ballet dancer, preparing for a theatrical show. Negulesco's frames are awash with color and movement: the twinkling lights of a matte-painted Eiffel Tower glimpsed through the window; the pinks, blues, greens, oranges, and chalky pastels of her fellow dancers' tops and tutus. A spiral staircase takes her down to an imagined stage on which she will dance. Jervis, glimpsed in a reaction shot sitting in a private seat, gazes at Julie with his binoculars. Throughout this sequence Astaire will be situated not as dancer but as viewer, the figure toward whom Julie's dancing and her desire are projected. Negulesco's framing throughout the Paris stretch of this sequence mostly keeps Caron's dancing body in full view, retreating back to long shots when she is joined by other dancers and cutting to closer shots during transitional moments. The camera's gaze, here, is not precisely Astaire's. Abandoning Astaire's seated position, the camera moves with Caron, and the film cuts on her movements. The mobile frame and the placement of the cuts are participants in the dance, partners in choreography. It is Caron who has broken Negulesco's frame out of aesthetic detachment and into body and energy, as Astaire, the diegetic viewer of this sequence, remains cloistered in his. Astaire sits quietly and, at the end of the number, applauds politely, amused but not overly impressed. Caron's dancing commands Negulesco's frame, but for Julie, Jervis remains out of reach.

The next sequence, in Hong Kong, pictures Caron as a cabaret dancer. Her movements exude a mature, uncharacteristic eroticism; the color of her orange dress matches the color of a jukebox that provides the score for this languorous dance. Julie tempts two American sailors, dances with two men, but her audience is again really Jervis, who sits, in a room apart from the main floor on which Julie dances, with two women. Astaire's relaxed posture, however, again suggests bemusement rather than substantive interest in Caron's dance. The frame keeps Caron mostly in the center, and in long shot, until a medium close-up near the end of the sequence, in which Caron, approaching Astaire, bids for his hand. But, suddenly, it is not Astaire. In a close reaction shot, he has been replaced by another actor. Jervis again remains out of reach, always able to slip away from Julie at the edges and corners of the wide frame. The final number, in Rio, finds Julie in a costume evocative of a Pierrot, reminding us of her French heritage. Astaire, rather than being figured as a viewer, now looms in the background while Caron dances, strolling across the set while the number proceeds, as Julie is joined by a troupe of dancers in outfits of various colors. The number includes three figures on stilts who evoke the charming drawing of "Daddy Long Legs" Caron conjured earlier in the film. Astaire, slipping away while his stilted doubles lumber about, eventually saunters offscreen.

Julie's dreams involve her trotting around the globe in search of Jervis; they are also variations on her own potential as an artist and a mature woman, and in bringing Caron to maturity also work to bring the narrative of *Daddy Long Legs* to a close as Julie discovers Jervis's identity as her benefactor. Even as this film moves toward its conclusion, however, Negulesco's widescreen cinema encourages us to linger with a pleasurable aesthetic gaze that takes priority over, even as it defers to and develops complex relationships with, narrative fulfillment. These two extended dance numbers in *Daddy Long Legs*—particularly Caron's ballet near the end of the film—are exemplary of Negulesco's taste for lingering on CinemaScope's aesthetic potential rather than its use as an instrument for immersive narrative drive. This aesthetic potential itself opens up, at least temporarily, glimpses of alternate ways of living that for a time also forestall the commodification of the wide image at play in the money-driven worlds of his films.

2

Blake Edwards (1922–2010)

Panavision Pyrotechnics

Films discussed in this chapter: *A Shot in the Dark* (1964); *The Party* (1968); *Darling Lili* (1970); *Wild Rovers* (1971); *"10"* (1979); *S.O.B.* (1981); *Victor/Victoria* (1982); *Skin Deep* (1989); *Switch* (1991).

Sitting Ducks

Blake Edwards was born into Hollywood. Edwards, né William Blake Crump, was the stepson of Jack McEdward, himself the son of Hollywood silent comedy director J. Gordon Edwards. Blake Edwards began as a bit player in films before going on to write two late classical Hollywood gems, *Drive a Crooked Road* (1954) and *My Sister Eileen* (1955), both directed by Edwards's important early collaborator, Richard Quine. The first two, and very delightful, films that Edwards directed—the musical *Bring Your Smile Along* (1955) and the gangster comedy *He Laughed Last* (1956)—demonstrate the director's aptitude for Technicolor in Academy aperture. Like Jean Negulesco, Edwards shortly thereafter became an early adopter of the colorful wide frame, using CinemaScope and Technicolor on both the noir *Mister Cory* (1957) and the comedy *This Happy Feeling* (1958). Perhaps the most surprising affinity between Edwards and Negulesco is that Edwards was also accomplished in the fine arts, although this talent was publicly exhibited only near the end of his life. Edwards enjoyed an exhibition of his artwork, a collection that includes both paintings and sculpture, in January 2009 at the Pacific Design Center in West Hollywood, California. These works express the comically spirited as well as colorfully abstract qualities also evident in his films. *Sitting Duck* (Figure 2.1), as the catalog accompanying the exhibition describes it, is "playful and whimsical in spirit" (*Art of Blake Edwards* 9) in its abstract conjuring of a befuddled duck, legs splayed on the ground. This duck's sculpted pose evokes a slapstick pratfall akin to those performed by Peter Sellers in Edwards's *The Pink Panther* (1963) and its sequels and *The Party*. Edwards's "Floating Space Series," meanwhile, is exuberant, with "a joyful, lyrical, 'jazzy' quality that sings in rhythmic progressions of color and shape" (13), words that could also describe many of Edwards's widescreen films, especially the ones scored by his career-long collaborator, composer Henry Mancini. The same exhibition catalog also describes the colorful abstractions at work in this series of paintings as forms that are frequently "gathering toward that central, sculptural figure in space" (13).

A variously centripetal and centrifugal quality also emerges in Edwards's play with the anamorphic frame, with the way his films place the human figure in a complexly unfolding and comic visual structure. John Falwell, analyzing the use of digression in *Skin Deep*, argues that

> Edwards recognizes the power of the film image to disrupt and disturb, and so makes even his most outlandish gags delicately abstract. . . . the gags have an underlying seriousness of intent that ties them strongly to the film's central drama. . . . Edwards' digressions allow him to do two things: first, to stray from his story the way a composer strays from his theme, avoiding the obvious path, teasing his audience with the pleasure of anticipated return, and testing his art to see if its structure can withstand the digression; second, the digression allows Edwards to accumulate a wealth of idiosyncratic material . . . along the way, lending his films the structure of a pilgrimage and the atmosphere of a carnival. (182)

Edwards's widescreen compositions often commence with a focus on an individual protagonist before activating surrounding areas of the anamorphic image outside of the purview of the central character. His wide frames will from that point eventually return, in the same or in subsequent shots, to a relatively narrower slice of the image that had been the erstwhile focus prior to the expansive activation of other parts of the frame. This aesthetic approach to widescreen complements the overriding theme that guides Edwards's career, which, as Peter Lehman and William Luhr write in the first volume of their two-volume study of the director, involves situations in which "Edwards's characters are continually confronted with the bizarre, with unexpected situations and events whose occurrences undermine any sense of rational order those characters may have

FIGURE 2.1 *Sitting Duck*, a sculpture by Blake Edwards. Author's collection.

previously had" (*Blake Edwards* 4). Dave Kehr ties this quality of Edwards's cinema more directly to the deployment of the anamorphic frame, noting how Edwards's "widescreen space becomes a vortex fraught with perils—hidden traps, aggressive objects, spaces that abruptly open *onto* other, unexpected spaces" (293). Chaos occasionally takes over Edwards's anamorphic frame but always in a manner underscored by the director's masterful visual orchestration of delectable disorder, a stylistic control that prefigures our eventual return to the character or narrative happenings that sparked the initial bedlam.

Early Anamorphic Diamonds: *A Shot in the Dark*, *The Party*, and *Wild Rovers*

Blake Edwards's early films found the director shifting back and forth between CinemaScope and nonanamorphic, 1.85:1 compositions. Despite his demonstration of fluency in the anamorphic canvas in early Technicolor films, Edwards during this period also utilizes the narrower 1.85:1 shape in *Operation Petticoat* (1959), *Breakfast at Tiffany's* (1961), *Experiment in Terror* (1962), and *Days of Wine and Roses* (1963), the last two films rare forays into black-and-white cinematography. After the success of the slapstick hit *The Pink Panther*, shot in Super Technirama 70, and its sequel *A Shot in the Dark*, filmed in Panavision, Edwards very rarely strays from the neighborhood of the 2.35:1 aspect ratio, apart from two later projects in 1.85:1—*Gunn* (1967), an adaptation of the television series *Peter Gunn* that Edwards created, projected in 1.85:1 theatrically, and his underrated remake of François Truffaut's *The Man Who Loved Women* (1983), also projected in a nonanamorphic "flat" format. These two later nonanamorphic films are made during the stretch of Edwards's career otherwise wholly committed to the anamorphic image.

A Shot in the Dark, the most perfectly realized of Edwards's *Pink Panther* films, is a good place to begin closely reading his work in the anamorphic frame. The film demonstrates his already refined ability at this point in his career to yoke the wide frame to existential themes.

A Shot in the Dark

A Shot in the Dark (shot by British cinematographer Christopher Challis, in his only collaboration with Edwards) is the second film in the *Pink Panther* series. It begins with a pre-credits sequence preparing viewers for the creative orchestration of the wide image in the film to come, and in ways that go beyond the mainly star-driven first *Panther* film, which activates expansive attention to the frame in isolated moments. Scored to Henry Mancini's "The Shadows of Paris," the sequence, set in a courtyard during late evening, finds the camera looking through the doors and windows of a mansion belonging to the Ballon family, headed by Monsieur Ballon (George Sanders). This opening is composed of elegant long takes collectively amounting to four minutes in duration. The interior lights of the various rooms illuminate the blues, grays, and greens of the mansion's architecture and the surrounding twinkle of the night sky. People are glimpsed furtively darting across the courtyard, covertly walking up and down stairs, quietly traversing hallways, stealthily slipping into rooms, and looking suspiciously out of windows. Each is making their way to a separate romantic dalliance that each, in turn, strives to keep secret within the purview of this wide image. They move as if they know someone is watching, and Edwards's camera—alternately tracking left and right and placed on a crane, prepared to move up and down as characters slip up and down staircases—is keenly positioned to follow. A note of comic irony is struck here in the anamorphic frame. Wide frames, which would seem to promise an expanse and a visual plenitude beyond narrower ratios, become instead in this opening sequence in *A Shot in the Dark* spaces where characters cloak and conceal themselves. The final such concealment in this opening sequence is the fatal action cited in the film's title: one figure walks into a room where two other figures are engaged in a romantic tryst. Edwards frames these figures behind walls and doors. Then, five shots are fired, the guilty party obscured. A cut to black takes us to the animated title sequence.

It will be up to Inspector Clouseau, played by Peter Sellers, to deduce—more precisely, stumble upon—the identity of the perpetrator. *A Shot in the Dark* creates a spirited tension between Clouseau's questionable methods of detection and Edwards's thoroughly assured widescreen compositions. As Daniel Varndell writes in his study of Sellers's performance in the first *Pink Panther*, "Peter Sellers perfects in his . . . Clouseau the look of a man whose unflinching self-belief is undercut by moments of exasperated self-awareness, whose unflappable dignity in the face of disaster registers, fleetingly, the barest flicker of consciousness at the vast absurdity of modern life" (136). In turn, Edwards's unwavering control of the anamorphic image reflects a careful orchestration of the absurdity and chaos impinging Clouseau's progress. Unlike Edwards's camera, which can anticipate where characters will go in the opening sequence of *A Shot in the Dark*, Sellers's Clouseau, although dimly aware, as Varndell suggests, of the absurdity swirling around him, is more often than not completely oblivious to the surrounding contents of the wide image revealed to viewers incrementally. And where the opening shots of the film implicitly acknowledge the inability of cinema, even an expansive cinema bolstered by widescreen technology, to see everything, Clouseau remains obstinate in his belief that he always has it within his complete power to grasp unambiguous truth and solve the case. The justifiable, partial omniscience of Edwards's wide frames is set into counterpoint in *A Shot in the Dark* with Clouseau's humorously unjustifiable confidence.

Clouseau's failure to thoroughly command the anamorphic frame as an authoritative narrative presence opens a vulnerable space in the wide image for violent gestures to intrude, resulting in fatal consequences for other characters. Late in the film, Clouseau romances a suspect in the case, Maria Gambrelli (Elke Sommer), hoping his dalliances with her will attract her jealous lover—whom Clouseau believes is the killer—to come out of hiding. This works, to a point:

Clouseau's gambit does attract the attention of a mysterious, shadowy figure, glimpsed only through the emergence of a black-gloved hand dipping into the edge of the frame repeatedly. It is eventually revealed, however, that this hidden figure is not the killer Clouseau seeks but rather the detective's supervisor, Commissioner Dreyfus (Herbert Lom), driven bonkers by Clouseau's incompetence. What is interesting in these sequences, which find Clouseau taking Maria to see flamenco performances, Hawaiian hula dancers, and Russian folk dances at various hot spots, is that Edwards begins each sequence by focusing on the performance within the frame before activating other areas of the image to draw the viewer's attention to the presence of a dangerous figure, stalking Clouseau with an eye to kill him, in the margins. Edwards slyly cleaves our engagement with the wide image from Clouseau's own point of view, a pattern throughout the film's use of the anamorphic image. Our efforts in comprehending details present onscreen are gently distanced from Clouseau's happy oblivion, our attention directed to other areas of the frame.

In the flamenco dance sequence, Edwards initially places his camera not from the point of view of Clouseau in the audience but alongside the guitar player and flamenco dancer onstage. The camera moves in rhythm with the music and the dancer's gestures, circling around the stage to eventually reveal the audience. The image is elegantly and vividly decorated by the bright red decor of the restaurant and the alluring movements of the flamenco dancers. But despite this pictorial pleasure, Edwards demonstrates a knack here for focusing our attention on sound's relationship to the expansive image. We are initially immersed in the music, and the rhythm of the scene is established by the intermittent sound of the dancer's shoes hitting the stage floor's hard wood. This immersion is rudely interrupted by the counterpoint of a very different sound: the low, monotonous, one-note drone of a nondiegetic synthesizer, heard as the killer's gloved hand emerges from the frame's right edge (Figure 2.2). The killer is seated in a private

FIGURE 2.2 *A Shot in the Dark* (United Artists, 1964). Digital frame enlargement.

box near the side of the stage, the area decorated by the fabric of a red curtain, as well as a table on which are arranged, laterally across the frame, a glass of champagne, glassware, and a candle. Although seated in a box with a perfect view of the performance, the hidden killer is placed at an angle opposite the stage, viewing not the dancers but Clouseau. The killer's gesture of putting out the light of the candle on the table, cloaking his presence in darkness, is unnecessary: Clouseau was already blissfully unaware of this gloved figure's presence. Our own awareness of the figure's presence offscreen is at odds with Clouseau's ignorance. The lateral expanse of the widescreen image that reveals the emerging hand engages our eye in a delectable fashion, again separating our way of seeing from Clouseau's own. The striking visual appeal of this shot—the sleek black gloved hand emerging from behind the curtain, the negative space on the left side of the frame that hints at the vulnerability of the eventual (offscreen and unintended) victim, and the bright red fabric of the curtain that pulls our eye past the gun that is revealed—only further emphasizes Clouseau's bumbling insensitivity to the vibrant details of his anamorphic world.

This scenario recurs in subsequent sequences, and in each case the killer's attempt to kill Clouseau fails, with a sudden change in the hapless inspector's positioning in each instance causing the killer to strike a quite innocent figure. The viewer remains the only witness of these murders. The knowledge we have gleaned from our widescreen-scanning engagement with gesture, composition, color, and movement, emotionally underscored and partially guided by the playful use of sound, is set into counterpoint against Clouseau's insensitive obliviousness. Clouseau's comic privilege in *A Shot in the Dark* is his blissful freedom to ignore everything at play around him in the anamorphic image.

The Party

In the sole non-*Panther* film Edwards made with Peter Sellers, *The Party*, the director situates the actor as a central point of visual interest, even as other figures, objects, and decor swirl about him. Edwards's cinematographer on *The Party* was Lucien Ballard, a film and television veteran who also worked with Sam Peckinpah on several productions. The film is notable for Edwards's use of on-set, video-assist technology in the composition of his images, a method Edwards employed following the lead of Jerry Lewis (who innovated the technology on his 1965 film *The Family Jewels*) (see Wasson 130), and which Edwards would continue to use subsequently. Edwards did not adopt this technique simply for efficiency. More consequentially in terms of *The Party*'s widescreen visual compositions, Edwards's use of video assist enabled him to compose shots that, at times, departed from the conventional framing of classical images. "*The Party* may be one of the most radically experimental films in Hollywood history," Lehman and Luhr note (*Blake Edwards* 163), precisely because the film is largely concerned with engaging the viewer's creative and contingent perception of the wide frame rather than using the wide image simply as a prefabricated vehicle for the delivery of narrative information. Slapstick gags instead draw attention to complex layers of staging, arrangement of objects, and movement across the frame, inviting explorative and playful rather than front-loaded, predetermined perception. The Panavision frame of *The Party* is a canvas the viewer must creatively work to scan, just as the Sellers figure struggles to physically maneuver the terrain of his world.

The Party follows Hrundi V. Bakshi (Sellers), a movie extra from India whose work in Hollywood is limited to appearing in the backgrounds of battle scenes in glossy, highly artificial widescreen Hollywood epics. (*The Party*'s innovative gesture is to throw this hapless, background innocent into anamorphic absurdity and chaos, a travesty of the epic, expensive, international coproductions fashionable during the period.) In the opening sequences he comically disrupts the filming of a spectacle that looks something like a latter-day *Gunga Din*, resulting in his expulsion from its set. Nevertheless, Bakshi remains tethered to the Hollywood community through being—mistakenly—invited to a swanky celebrity party thrown by huffy, self-important film producer C. S. Divot (Gavin MacLeod) at the home of a studio head, a former military general named Clutterbuck (J. Edward McKinley) and his wife, society woman Molly Clutterbuck (Kathe Greene). The film's ironic attitude toward the artifice of commercial Hollywood spectacle (an irony signaled in the casting of Sellers to play a character from India in the first place, and prefiguring the more acrid attitude toward Hollywood in Edwards's later *S.O.B.*) situates Hrundi as a marginalized figure within what the film presents as the repressive, militaristic mode of Hollywood film production. *The Party* itself, in counterpoint to the epics travestied by this opening sequence, uses the film image as a site for compositional play and carefully improvised and staged gags, rather than for authentic landscapes or grand vistas.

In this way, *The Party* works in a register slightly different from Jacques Tati's contemporaneous *Play Time* (1967), the aesthetic similarities between the two films notwithstanding. Tati is largely concerned, as Malcolm Turvey puts it, with "the dissolution of the comedian" in the context of French modernity, given that his trademark character, Monsieur Hulot, is offscreen for nearly half of *Play Time*'s duration. "This privileging of the comedian is compounded by the lack of information about Hulot's mental state" (80) in *Play Time* and other Tati films, opening up Tati's work to what Turvey (following Tati's own notion) calls a kind of "comic democracy," wherein the "right of everyone who appears on the screen to be funny" is acknowledged (Stéphane Goudet, qtd. in Turvey 53). While there are many funny supporting figures in *The Party*, Sellers remains onscreen in the majority of the film's scenes, and the film's humor is defined in relation to him.

The Party's "comic democracy," however, is felt on another level, not so much performatively—in the sense that anyone in the film can be quite as funny as Sellers (they cannot)—but compositionally, in terms of Edwards's play with the wide image. Sellers's movement through *The Party* acts as an implicit tutorial for the viewer's own visual engagement. By watching Sellers, we become engaged in a widescreen guessing game, our eyes working in concert with the actor's gestures and movements not in order to detect some kind of depth of psychological interiority (there is none) but rather to explore the nooks and crannies of a complexly staged and mounted proscenium.

The first extended gag sequence in the Clutterbuck home tutors the viewer in how to perceive and anticipate the activations of wide Edwardian space. Upon arriving at the home, Sellers becomes aware he has dirt on his shoe. Not wishing to track footprints all over the modern, angular home of the Clutterbucks, Hrundi dips his shoe into one part of the home's fountain, a complex system of flowing water circulating throughout the house. The shoe is cleaned, but it slips off Sellers's foot, floating down the pool and through a drain. From there, it bobs and floats to a lower part of the fountain in the central area of the home, where socially important guests hobnob. To retrieve his shoe, Sellers must not only clandestinely dart his hesitantly puckish way through this aqua-maze but also navigate his way around other figures emerging in the frame—including a jazz band whose entrance to the Clutterbuck home sweeps Sellers up and carries him toward where other guests are chatting. These images of the Clutterbuck home emphasize its sleek, angular lines, underscored by the careful, frequently symmetrical, lateral placement of props and objects across the mise-en-scène. But whenever it appears, the perpetually flowing water operates in contradistinction to this smoothly arranged surface, the water's flow ceaselessly guiding our eye out of frame, to the next slice of screen space across which we discover more and more of the Clutterbuck home and within which Sellers—in tandem with Edwards—devises new strategies for navigating from one side of the frame to the other, and across the depth of the frame.

Sellers repurposes, first, a paper party invitation, enabling him to walk across to the other side of the fountain without other guests noticing his missing shoe; and, then, a plastic plant, which he uses to try to fish out the missing footwear, presently stuck in between two stepping stones in the middle of the interior pool. This moment is less about our perception of the shoe's precarious position in the water, however, and more about establishing the film's intent, via Sellers, to use objects to guide our eye from one frame to the other, and within and across single shots. Sellers has hooked the shoe on a plant limb, like a fish on a line, but in attempting to "reel in" the shoe, he accidentally flings it across the room, through the open door of the kitchen, after being distracted by a waiter (Steve Franken) who as the film goes on becomes increasingly drunk on beverages he is supposed to be serving to guests. The flung shoe ends up on a tray of hors d'oeuvres, brought to various guests who do not seem to notice the footwear adorning it. When the waiter reaches Sellers, the shoe is returned to its owner, the sequence culminating as if in a secret aesthetic logic. This sequence with the shoe demonstrates the variously centripetal and centrifugal aspects of Edwards's and Sellers's work within the widescreen frame in *The Party*.

Wild Rovers

If *A Shot in the Dark* and *The Party* demonstrate Edwards's interest in exploring the anamorphic image in relation to the human figure, *Wild Rovers*, his only western, complements it with an equally intriguing play with scale across the cuts joining the widescreen shots in the film.[1] The cinematographer on *Wild Rovers* was Philip H. Lathrop, who in addition to many episodes of *Peter Gunn* and *Mr. Lucky* also worked with Edwards on the earlier features *The Perfect Furlough*, *Breakfast at Tiffany's*, *Experiment in Terror*, *Days of Wine and Roses*, *The Pink Panther*, *What Did You Do in the War, Daddy?*, and *Gunn*. *Wild Rovers* is about two cowboys—Ross Bodine (William Holden) and Frank Post (Ryan O'Neal)—who take a stab at life outside the norms of conventional, wage-labor work. Their plan: rob a bank, flee to Mexico, and live free. The outward simplicity of this scheme is thrown into relief against the complexity of the anamorphic frames across which they rove. As previous commentators have noted, *Wild Rovers* dedramatizes, even as it still includes, many of the archetypal narrative events of the genre—bank robbery, shootouts, card games, bronco busting—in a style reveling in the playful possibilities of framing and scale in the genre (see Lehman and Luhr, *Blake Edwards* 195).

Wild Rovers dispenses with the usual parallel in the western genre between man and nature, a departure key to its exploration of the Bodine-Post relationship in the context of the widescreen frame. As Harper Cossar observes in his discussion of the CinemaScope style of Otto Preminger's *River of No Return* (1954), "Preminger foregrounds the frame's dimensions, but he does so to equate man with nature (a western generic trope): Matt is of equal stature to the tree he is felling" (103). In the Panavision frames of *Wild Rovers*, Edwards repeatedly underscores that his men are *not* of equal stature to nature. It is not that nature overwhelms them, although this initially seems implied in the film's opening credits sequence, in which Holden and O'Neal ride, mainly in silhouette, across landscapes at dawn. Nature, rather, stands autonomously from them. Edwards plays with this idea in the film's sudden shifts in scale in the framing of nature, which frequently involves a shifting scale in the film's representation of animals in the wide image. When we first meet Bodine and Post, they are working as ranch hands for Walt Buckman (Karl Malden). As the men prepare for work, mounting horses, Edwards does not show them as visual equivalents to the horses they are riding. The body of a horse instead offers an opportunity for alternating scale in the frame. At certain points nearly the entirety of the anamorphic image is engulfed, during this first ranch sequence, by partial views of

[1] Edwards has been interested in the western genre from an early point in his career. An episode from the first season of *Peter Gunn*, entitled "Pecos Pete" (initial airdate February 9, 1959), and with a story by Edwards, transposes the urban private detective of the title to a Texas setting full of western iconography.

the bodies of the horses, as in one shot that, showing the men getting ready to go, focuses our attention on several horse hooves kicking up dust. This shot gives way eventually to a series of closer shots of the men, against the blue sky, preparing to go to work. The film's title card is presented against a shot of the men riding the horses toward the cattle pens, and taken from a low angle the horses seem imposing figures. Frequently Edwards will abstract his close-ups of the men from the horses, suggesting not a parallel between man and nature but rather a tension between them.

Horses are not the only animals in *Wild Rovers* subject to unconventional shifts of visual scale. Dogs and cats appear nearly as often, and dominate the volume of the frame in alternately charming and startling ways. Often these animals occupy a space in Edwards's imagery usually occupied in this genre by more familiar elements, such as a gun or a brawl. Bodine and Post begin their robbery at the home of a banker, Joe Billings (James Olson); while Bodine and Billings go to the bank to retrieve the money, Post is charged with keeping an eye on Billings's wife and daughter. As played by Ryan O'Neal, Post is alertly sensitive and, for a bank robber, surprisingly unaggressive, as comfortable in this domestic abode as he is in the Wild West, a fact expressed in this scene through the way he places his gun on top of a book (framed very close to the camera on the left side of the frame, the book's place in the mise-en-scène more voluminous than the gun itself) in order to pet a dog with both hands. In Edwards's cutting, the way in which the dog occupies the center of the frame—in O'Neal's arms—emphasizes how this gentle animal replaces the weapon previously wielded. After Bodine and Billings return with the money, things go awry: a threatening leopard has attacked the Billings home while they are away—killing the poor dog—and Edwards amplifies the emotion of the moment by eventually revealing to us that this dog had puppies, now without a mother. It is a startling and surprising occurrence given the large cat's sudden presence has not been foretold through any kind of causal chain. The sudden appearance of the imposing feline in close-up affords Edwards the opportunity to play with visual scale across shots.

In a previous shot, the final image in a sequence that sees Bodine and Billings ride out of town after stealing the money from the bank, Edwards frames from a long-shot distance a boisterous brawl outside of a cathouse. At the end of this sequence, there is a cut back to the Billings home, and the first shot is of the threatening leopard, its visage engulfing the frame. Edwards does not rely only on the sudden presence of the leopard to generate surprise; he uses unexpected variations in volume from shot to shot to create the jolt. Later, shortly before Bodine and Post stop to get a horse from a friend of Bodine's, Ben (Moses Gunn), it is revealed that Post has swiped one of the pups from the family home. The puppy's minuscule presence in the frame in relation to the two cowboys is charming and comic, and another reminder that the vistas of *Wild Rovers* constantly underscore the otherness of animals and nature in relation to the men. Bodine eventually persuades Post to give up the pup (to Ben, who will provide in exchange a horse for them to ride). When last seen, the puppy is shown by Edwards joining a litter of kittens to suckle at the teat of a mother cat, the latter effectively replacing the dog's dead mother. This mother cat and the members of its litter occupy the entire wide image just as the leopard had done previously, but now the effect is touching rather than terrifying.

Edwards returns, late in this film, to the idea of an animal—again a horse—as a creature autonomous from man, once more through a play with scale and figuration in the frame. On their way to Mexico, Bodine and Post encounter a wild bronco that they seek to bust, to replace the stubborn donkey they have been stuck with during this part of their journey. In the ensuing narrative moment, Bodine does successfully tame the bronco, at least enough to ride it. The very nature of the landscape in this bronco-busting sequence is unusual: rather than giving us a vision of the dusty, rocky, tumbleweed-strewn, desaturated prairie we might expect in a widescreen western, Edwards gives us a flatland covered in vivid white snow. And he treats us to several images emphasizing the size and speed of the horse, images that diminish Holden and O'Neal—two figures dragged by the animal along the frozen ground. Even frames that would appear to balance the composition of the Scope frame between man and animal instead stress the ornery resistance of the horse to Holden's efforts to corral it, rendering ironic any equivalency implicit in the composition of the shot.

Even more intriguing is the final stretch of the sequence, as O'Neal watches with pleasure as Holden successfully rides the bronco for the first time. O'Neal is presented alone, against the frozen landscape, cheering Holden on as he rides the beast. Before too long, Edwards renders these images of both men, and the horse, in slow motion, using lap dissolves to superimpose close-ups of Holden riding the bucking bronco with shots of O'Neal leaping in the air, vicariously enjoying Holden's taming of the animal. These lap dissolves further relativize the scale between man and nature across the anamorphic frames of *Wild Rovers*. At one moment Edwards shows us the hooves of a horse kicking up snow, superimposed and in slow motion over the image of O'Neal twirling about in celebration; in another, he will show Holden riding the bucking horse (with O'Neal giving chase on the left side of the frame), Holden's bucking gesture superimposed over the same gesture in close-up (Figure 2.3); in yet another he shows us Holden, in close-up, riding the bronco while O'Neal, in superimposed long shot, stands on his head, a juxtaposition of Post's vicarious celebration of Bodine's masculinity with the genuine article. These heterogeneous widescreen images creatively trouble any coherent or predetermined sense of what a Scope western, or indeed a cowboy, should look like.

Three with Julie Andrews: *Darling Lili*, *S.O.B.*, and *Victor/Victoria*

In contrast to the comically irrational poise struck by Peter Sellers in his work with Edwards, in which the filmmaker's control of the frame acts in

FIGURE 2.3 *Wild Rovers* (Metro-Goldwyn-Mayer, 1971). Digital frame enlargement.

counterpoint to the chaos swirling about the Sellers figure, and in further contrast to the focus on a largely masculine world in *Wild Rovers*, Julie Andrews is a figure who motivates different forms of expansion in and activations of Edwards's widescreen frame.

The Flowers of *Darling Lili*

In *Darling Lili*, Lili Schmidt (Andrews) is, like the actress who plays her, a brilliantly talented singer and performer. Lili is also a German spy in World War I. Her mission is to extract sensitive information about military plans from an American fighter pilot, Major William Larabee (Rock Hudson). While in France she doubles as Lili Smith, musical entertainer of Allied troops, her German identity (Lili is half German, and half British by birth) cloaked behind an English moniker. With Larabee, however, she becomes romantically interested and, eventually, involved, complicating her hold on a fabricated identity. Larabee's courtship of Lili is central to the first half of the film, his seduction largely dependent on his ability to decorate the mise-en-scène around which he encounters her. Lili, for her part, will need to maintain her duplicitous identity when she is around Larabee, even as the wide frame around her introduces unpredictable contingencies. The romantically inclined Larabee will bring a group of violinists to Lili's window, luring her out to the grounds below for a candlelit picnic under the moonlight. Apart from the music, this sequence is silent—Andrews and Hudson speak, but we do not hear their words, the sound of the musicians commanding the soundtrack. The sequence contains a series of stylistic gestures by Edwards, organized around the intertwined presence of flowers and Julie Andrews.

The moonlit picnic sequence begins and ends with the use of the zoom lens, a technique of mobile framing that enables Edwards and cinematographer Russell Harlan (a previous collaborator with Edwards on *Operation Petticoat* and *The Great Race*) to explore relative and shifting scales of imagery in the wide frame.

As we have seen earlier in this chapter, in *Wild Rovers*, made a year after *Darling Lili*, Edwards deploys unconventional editing patterns to disrupt relations of scale across frames of different shot distances. Such manipulation of the wide frame implies the presence of an editor, juxtaposing expansive images of various scale against one another for an effect achieved at some distance from the filming of the scenes themselves. In *Darling Lili*, by contrast, the intervention of the zoom lens into the wide image suggests that there is another subjective consciousness, a third party in addition to Lili and Larabee, implicated in the romantic dalliances between the two characters and living alongside them at the moment of filming. It would not be a stretch to read this subjectivity as a self-projection of Edwards himself into the wide frame, given that *Darling Lili* is a romance that Edwards scripted expressly for Andrews, whom he would marry during the course of this film's postproduction. In this picnic sequence, Edwards begins with a zoom-out, his camera attentive to a batch of daisies gathered in a vase in the center of the picnic spread. As the camera zooms out, this close-up of the daisies gradually becomes a long shot of Andrews and Hudson seated across the picnic spread, with the musicians and candlelight surrounding them. This adjustment of the zoom lens also reveals the apparent placement of the camera, not near the daisies, as it might be perceived to have been, but rather from behind a tree. Two of its branches diagonally traverse the image, forming an internal frame. The wide image in this way becomes a tricky and initially misleading imbrication of both intimacy and distance.

And yet because no stable, permanent position is ever possible in Edwards's variously centripetal and centrifugal play with the anamorphic frame, our view of this picnic will suddenly change. In a creative violation of a textbook rule about mobile framing—that the movement of the frame must come to a complete stop before the editor cuts to the subsequent shot—Edwards cuts away from this shot behind the tree before the gradual zoom-out is completed. The effect of this cutting-on-frame-movement is a sudden shift of scale not unlike that described in *Wild Rovers*, but which in *Darling Lili* takes on different connotations. Because the shot Edwards is cutting to here is of Julie Andrews, the effect of transitioning to a close-up, from what was previously a long-shot distance, only intensifies our sense of an intimate attachment to a figure in the anamorphic frame. It is almost as if Edwards is seeking to restore, through this cut, the initial closeness of the frame to the picnic with which the sequence began, with Andrews now serving as an even greater point of interest than the daisies. Andrews's gestures within the frame will now direct our attention to the ways in which the actress is associated with these flowers. From the close-up of Andrews, Edwards cuts to a close shot of Hudson and then to a series of medium shots of Andrews and Hudson enjoying their picnic. It is in the second and third of these where Andrews's gestures once again draw our attention to the daisies. She plucks a daisy from the bunch, smells it, and then dips the flower into her glass of champagne, twirling its petals in the liquid and then smiling at Hudson. This gesture cues another series of close-ups exchanged between Hudson and Andrews. But apart from the music, this entire sequence is silent, drawing our attention away from whatever Andrews might be saying to Hudson and back to her gesture with the flowers, which becomes the main performative substance of her work in Edwards's image.

That the romance between Lili and Larabee—the main narrative content represented here—is now advanced further, is more than clear enough. And so Edwards ends not with the exchange of glances between the two diegetic lovers but with an idiosyncratic presence in the frame, tied again to Edwards's use of the zoom lens. Rather than zoom out, as he did at the beginning of the sequence, Edwards now zooms in, and this time not to a bundle of daisies but rather to a single one tucked behind the ear of the waiter Larabee has employed to attend to this elegant picnic, seated behind the couple. The flower in his ear is visible as a background detail earlier in the scene, even as our attention was more centrally focused on Andrews and Hudson. But now the zoom-in to this flower becomes the detail in the frame with which the sequence will conclude, a bookend to the initial zoom-out from the daisies. We might also note here that the waiter, as he gazes at a violin player, is crying. That he is staring at a male musician as he weeps adds texture to the scene, one that implies the possibility of other ways of connecting with the romantic courtship at play in these anamorphic images.

Lili continues to dote on daisies, twirling one between her fingers during a later scene, a meal with her German compatriot, Colonel Kurt Von Ruger (Jeremy Kemp), a sequence that begins with a rack focus shot of a daisy in which the flower momentarily takes up the entire expanse of the frame. The flower motif then continues as Larabee, in a subsequent scene, has delivered to her a gift of more flowers, now a large batch of roses. But now, rather than occupying only a part of the frame, as the daisies did in the picnic sequence and in the sequence with the Colonel, the roses will command a greater and greater share of the expanse of the image. Initially, Larabee's gift appears to take the form of only a single rose, accompanied by a note, delivered to Lili while she is still sleeping. The length of this single rose, its petals resting on a pillow next to Andrews's head, takes up most of the lateral stretch of the image. Lili awakes and sees the rose and the accompanying note. "One rose, one tender thought of you," the note from Larabee reads in close-up. Edwards will again implicate his presence in the frame through a cut from the note not to a reaction shot of Andrews reading it as we might expect but rather to an extreme close-up of another rose. This flower is delicately placed in focus in the foreground of the frame, its red petals and green stem thrown into relief in this rack focus shot against smeary greens and reds—which will soon be revealed to be other roses—in the background of the shot. Edwards, in this complex image, matches and exceeds Larabee's gift of a single rose with his cinematic revelation of many. Without cutting, his rack focus shifts from an emphasis on the single rose to the several lying behind it, and then, as the camera tilts upward, to Andrews herself, internally framed by the petals and

FIGURE 2.4 *Darling Lili* (Paramount Pictures, 1970). Digital frame enlargement.

stems of the roses in front of the camera as the focus adjusts again to draw her into clear view. At her gaze, the shot then zooms out as the camera also tracks, describing a semicircle as it follows Andrews as she walks among many dozens and dozens of roses arranged across the room (see Figure 2.4 for a juxtaposition of the beginning (top) and ending (bottom) frames of this complex image).

In the narrative world of *Darling Lili*, this splendid arrangement of beautiful roses is the work of Larabee. Cinematically, these roses placed across and within the anamorphic widescreen frame are the work of Blake Edwards, the achievement with the roses one of widescreen technique in tune with Julie Andrews's gestures, movements, and expressions. This intermingling of Edwards's projection of his authorial presence in the wide frame with Larabee's romantic subjectivity can of course be read as the auteur's implication of self in a romantic narrative involving the woman he was to marry during the production of the film. But given that the uncertainty of socially legible identity is the main theme of *Darling Lili*, such gestures of style also generate ambiguities about sexuality explored further in the director's later work in the anamorphic image.

S.O.B.

In comparison to the lush romanticism of *Darling Lili*, *S.O.B.*, a collaboration between Edwards and cinematographer Harry Stradling Jr. (who would work again with Edwards later in the 1980s on *Micki + Maude*, *A Fine Mess*, and *Blind Date*), is marked by a thorough distaste for the ways Hollywood's commercial imperatives disrupt the creation of cinematic art. The screenplay to *S.O.B.* was written by Edwards in the early 1970s, a difficult period when the director was reeling from the financial failure of *Darling Lili* and studio interference in the making of both *Wild Rovers* and *The Carey Treatment* (1972). Edwards's bitterness toward Hollywood studios in light of the fate of these earlier films is palpable in *S.O.B.*, in which a jaded Hollywood producer, Felix Farmer (Robert Mulligan), cynically remakes a traditional family movie musical, *Night Wind*, into a soft-core sex epic after the film's first, wholesome iteration fails commercially. Andrews plays Farmer's wife, Sally Miles, an actress and singer with whom Farmer has made several previous films, including the first version of *Night Wind*. Farmer's scheme in remaking the picture involves a complete reimagining of *Night Wind*, and a reimagining of Sally's earlier character, transformed from an asexual, Peter Pan–like innocent into an adult women possessed of sexual desire.

Farmer's reshaping of Sally Miles's image in *S.O.B.* resonates with the offscreen, intertwined lives of Edwards and Andrews. Frequently noted is that in her collaborations with Edwards, Andrews reshaped perceptions of her relatively wholesome, family-friendly image, established in films such as *Mary Poppins* (1964) and *The Sound of Music* (1965). This remodeling of Andrews's star persona is prefigured in *Darling Lili*. In the first sequence of that film, Lili performs a polite musical number for troops in a musical hall; midway through the film she is performing a steamier burlesque number in a nightclub. In both sequences Edwards's camera is intimately attuned to Andrews's performance, first in the form of a complex long take in the opening music hall performance, via a camera tracking around the stage alongside Andrews, and then via relatively more frenzied cutting patterns during the burlesque. Edwards's wide frame becomes her dance partner in these sequences, the choreography of the compositions and placements of the cuts working in concert with Andrews. But where the coupling of auteur and star in *Darling Lili* bequeaths the anamorphic frame a touching, intimate romanticism, *S.O.B.* infuses that frame with palpable cynicism generated by a studio system's interference with projects like *Darling Lili*. Edwards is in this way still implicating his subjectivity as an artist through his use of the widescreen frame in *S.O.B.*, but it is to an altogether different purpose than in *Darling Lili*. The anamorphic frame in *S.O.B.*, far from providing an ostensibly expansive space within which affection or romantic intimacy is carved out between Edwards and Andrews, functions in a bitterly and comically ironic way, presenting the relationship between the filmmaker and actress characters in

S.O.B. as tainted by crass commerciality. Andrews's performance is particularly important to *S.O.B.*'s use of the wide frame in two key scenes. In these sequences, Edwards's subjectivity as author is intermixed with the collective subjectivity of the filmmakers in *S.O.B.*, the frames of the film implicating Edwards in the operations of the industry he is critiquing.

The first of these sequences is a musical number, a representation of Felix Farmer's *Night Wind* in its first, family-friendly incarnation. The subjectivity of Edwards the auteur and that of Farmer, the diegetic producer of *Night Wind*, are here enmeshed: the film we are seeing is *S.O.B.*, but within the context of the fiction it is also *Night Wind*. And like *S.O.B.*, *Night Wind* is also shot in anamorphic widescreen. Later sequences in *S.O.B.*, which show the characters watching reels from *Night Wind* in a projection room fitted with a Scope screen, confirm the aspect ratio of Farmer's film as more or less matching Edwards's own. But the space of *Night Wind* represented in these anamorphic frames is a cinematic abstraction, of no contextual significance besides serving as evidence of the existence of *Night Wind* within *S.O.B.* itself. And the fact that Felix Farmer—in *S.O.B.*'s world, a producer rather than a director—is eventually situated as the "auteur" of these images suggests what we are seeing is manifestly impersonal and commercial. The frames in this studio-set sequence are decorated with oversize objects representing childhood (a gigantic jack-in-the-box, a cannon that shoots confetti, letter blocks, colorful balloons) and costumed personages populating this child's world: toy soldiers, three dancing ballerinas, and a giant bear, among other buoyant figures. Edwards's camera follows Andrews's movement across this playful space, only occasionally cutting in for closer shots as she navigates *Night Wind*'s dreamlike world. The climax of her performance is matched by the crescendo of Edwards's swooping, upward-moving crane shot, as Andrews dances with all these plush representations of childhood memories. Here the wide frame becomes a toy box proscenium (Figure 2.5).

FIGURE 2.5 *S.O.B.* (Paramount Pictures, 1981). Digital frame enlargement.

Given the tendency of Edwards to keep the camera at some distance from all this frolicking, at no point does this sequence from *Night Wind* invite escape into another reality, even as the toy box world of the anamorphic image seems to offer infantile immersion. The sequence could in fact be received as a lighthearted parody of early Julie Andrews musicals. But while both Andrews the star and Edwards the auteur are implicated parodically in the frames of this opening number, these implications are more oblique, and more cynical, than those in the flower sequences of *Darling Lili*. The arrangement of childhood toys and the brief use of animation combined with live action—in a medium close-up shot in which a group of colorful balloons begin singing with Andrews—recalls the innocent, childlike image Andrews projects alongside similarly animated figures in *Mary Poppins*. And the creative use of the widescreen frame is reminiscent of any number of Edwards's initial experiments in widescreen, from the play with color and shape in his early college comedy *High Time* (1960) to the unhinged slapstick fantasy of the later *Pink Panther* sequels Edwards made with Peter Sellers (*The Return of the Pink Panther*, 1975; *The Pink Panther Strikes Again*, 1976; and *Revenge of the Pink Panther*, 1978). And yet the placement of all this whimsy in the toy box space depicted in the opening sequence of *S.O.B.* throws a question mark around its contents, here abstracted not only from the narrative of *S.O.B.* itself—which has yet to begin—but also from the diegetic film *Night Wind*, about which we never learn very much. To the extent that it functions as a representation of the earnest intentions of the makers of *Night Wind*, this opening sequence is not expressive of any kind of personal statement. Felix Farmer does not want to remake his movie because it has failed to achieve personal, aesthetic goals. He wants to remake it because his film has lost an incredible amount of money.

When Farmer comes up with his idea to rescue *Night Wind* from commercial oblivion by transforming it into a soft-core sex epic, he describes reshoots that will reimagine and recontextualize everything in the film he made. Just about all we eventually see of this reimagining in *S.O.B.* is a new version of the aforementioned toy box sequence. Now, instead of playing an innocent, childlike character frolicking with toys, Andrews, and Sally Miles, will play a woman tortured by visions of her unbridled, adult sexuality. This sequence notoriously involved Andrews exposing her breasts to the camera, a notoriety that resonates within the narrative world of *S.O.B.*, since it is also implies that the character Sally Miles possesses a similarly "innocent" public image. What is interesting about this reimagining of *Night Wind*, at least in terms of how it is represented by Edwards as a part of *S.O.B.* and as a collaboration with Andrews, is that this baring of the performative self in front of a director who is also the performing self's husband actually lacks the intimate gestures of the zoom lens and rack focus glimpsed in the earlier anamorphic frames of *Darling Lili*. But at the same time, the sequence is also distinctive from the surrealistically detached toy box sequence that opens *S.O.B.* Edwards's camera is more involved in the spatial terrain of this subsequent reimagining of *Night Wind*; we never get the complete proscenium of space that we get in the first version of the toy box sequence. This is a fragmented, nightmarish reimagining of the opening number that activates the anamorphic frame in an entirely different way.

Where the toy box sequence from *Night Wind* was presented to us in its finished form, the reimagined version of this sequence in *S.O.B.* is bookended by shots of Felix Farmer's film crew creating the sequence we are presently watching. As with the opening sequence, but in a now more explicit way, we see an intersubjective mix of Edwards-the-auteur crafting the wide frame, and the film crew of *Night Wind* assembling their crass reimagining of their film. The sequence begins with a shot of Sally Miles standing in front of a carboard set representing Satan's mouth and teeth. She is wearing a red dress, matched by the red of the devil and by the color filters inflecting the fog swirling inside Satan's mouth. Edwards's framings in the opening toy box sequence established a spatially coherent proscenium organized laterally across the frame, with only occasional cut-ins to Andrews. By contrast, this reimagining of *Night Wind* traffics in intentionally incoherent spatiality. No coherent proscenium establishing spatial relationships is conveyed in the reimagined version. The camera follows Andrews as she walks inside the devil's mouth, beckoned inside by a shirtless, muscular man, with a strong facial profile reminiscent of Edwards himself. A cut to a long shot then takes us inside the mouth, as Andrews emerges from a single red point in the middle of the frame amid an expanse of blackness. After a moment, the camera cuts in to closer shots of Andrews as her image is mirrored around her, in a style that suggests an anamorphic reimagining of the funhouse sequence at the end of Orson Welles's Academy ratio *The Lady of Shanghai* (1948), as well as a variation of Edwards's own use of mirrors in a climactic sequence in the earlier *Gunn*. From here, the camera and the gradually revealed mirrors multiply the images of both Sally and the dancer who is beckoning her into a mysterious world. This representation of the devil's mouth, and of adult desire, as a scary funhouse of mirrors that swallows up lateral visual expanse insists on the multiplication of Sally's image as Edwards's camera tracks backward to reveal more of the space (ultimately, a revelation of more mirrors). This sequence invokes earlier uses of the CinemaScope format, in the 1950s, as a means of ostentatiously displaying the exposed female body (in, especially, Negulesco's *How to Marry a Millionaire*), an invocation that is heightened by Andrews's exposure of her breasts at the end of the sequence, a gesture that is reflected in the mirrors placed across the widescreen image. But at every point the sequence insists on the fragmentation of Andrews's image in the frame, as if to suggest that this spectacular presentation of the female body comes at the cost of a coherent proscenium. This devilish reimagining of *Night Wind* does certainly break free of the childish repression of the opening toy box sequence, reclaiming the anamorphic frame as a space for adult sexuality and psychology. But as the cutaway shot to a comically startled film crew upon Andrews's disrobing reminds us, what we are to make of this gesture of apparent liberation in the cynical and commercially tainted world of *S.O.B.* is far from certain.

Victor/Victoria

Edwards and Andrews return to the musical form with *Victor/Victoria* (shot by cinematographer Dick Bush, a frequent Edwards collaborator who will return to work with the director on *Trail of the Pink Panther*, *Curse of the Pink Panther*, *Switch*, and the short-lived 1992 TV series Edwards created with Julie Andrews, *Julie*). The film is a remake of the 1933 German film *Viktor und Viktoria*. Both films tell the story of a singer who masquerades as a man when she is unable to find work as a traditional female performer. In the Edwards version, Victoria Grant (Julie Andrews) transforms herself into the nightclub sensation Victor Grazinski, a Polish count. Victoria, struggling to find work as a singer during the depths of the Great Depression, successfully assumes the identity of a wealthy male aristocrat in her nightclub performance. Her characterization of Victor is guided by the mentorship of Carroll "Toddy" Todd (Robert Preston), a gay singer and performer similarly down on his luck when he is fired from his job as an entertainer at the Chez Lui in Paris. A Chicago gangster, King Marchand (James Garland), and his girlfriend, ditzy moll Norma Cassidy (Lesley Ann Warren), are both enchanted by the talents of Count Grazinski, now one of the popular singers in Paris. But Marchand, who falls in love with the Count—and remaining steadfast, in his sexual insecurity, that the she pretending to be a he is in fact a she, even when she herself remains adamant that he is a he—is disturbed to find that she is, apparently, a he. Much of the film involves Victoria performing both privately and publicly: convincing her audiences in the film's world of her identity as Victor in her performance, and also maintaining her ruse as Victor in her dalliances with Marchand.

Edwards's play with the frame in *Victor/Victoria* beautifully orchestrates this theme of the performance of private and public identities. The film is replete with internal frames within the wider anamorphic frame that variously bracket or blend private, or publicly undisclosed, forms of being. These internal frames take various and fluid forms. Windows, mirrors, doors, curtains, and vertical lines of all sorts mark out interior borders, forming private frames within larger public spaces that afford insight into character psychology and personality. However, the internal frames in *Victor/Victoria* do not serve as "authentic" depictions of private selves in relation to the inauthenticity of public, social performance in the wider frame (see Lehman and Luhr, *Blake Edwards: Returning to the Scene* 50). Instead, Edwards, through his organization of the frames in *Victor/Victoria*, suggests private self and public appearance are both performances that perpetually shape and reshape identity in relation to an unstable and complex public world.

Edwards establishes this relationship between internal frames and the world of the larger anamorphic frame in the first scene. On a snowy Paris morning, Toddy sleeps next to a lover, Richard Di Nardo (Malcolm Jamieson). The camera begins on a shot of the Paris streets outside Toddy's window and then tracks backward to reveal his apartment. As the tracking shot continues, a small poster

FIGURE 2.6 *Victor/Victoria* (Metro-Goldwyn-Mayer/United Artists, 1982). Digital frame enlargement.

of Marlene Dietrich is revealed, suggesting the film's indebtedness to German film history and immediately associating the film's world with cross-dressing, queer sexuality, and sexual ambiguity. The camera eventually lands on Toddy, sleeping. Richard gets up and prepares to leave, hustling Toddy for money. He walks over to the dresser and opens it, in doing so repositioning the mirror hanging on the left door so that Toddy, now awake, is visible in its reflection (Figure 2.6). Initially, the reflection is framed internally within the mirror itself, as Richard gathers his clothing on the left side of the frame. After a moment, Edwards tracks the camera toward the mirror, the borders of which gradually disappear as Toddy's mirrored reflection takes up the entirety of the anamorphic frame as he quotes Shakespeare—"therefore is winged Cupid painted blind"—to Richard. Toddy's words suggest the ambiguities of love and sex (with the implication that he has loved Richard on some level while Richard is only in it for the money). Edwards's play with internal frames suggests their own ambiguities, as Toddy's very image within the wider frame shifts from a reflection of a private self (of Toddy, alone, in bed) to a potentially public one, as the mirrored, reflective image becomes *the only image* during the few seconds in which it commands the entirety of the wide frame, the mirror's borders kept momentarily offscreen. An internal, private frame has become, momentarily, an external—potentially public—one, setting into visual motion the film's exploration of private and public worlds.

The first appearance of Victoria Grant also involves an internal frame. After a failed audition, Victoria, poor in spirit and money, wanders the snowy Paris streets, happening upon a French bistro. The camera, positioned inside the establishment, is on one side of a rotund man who lavishly enjoys a frosted pastry. Behind him is a window through which the streets Victoria wanders are visible, and in a moment she enters this internal frame. The large man, positioned closer to the camera, takes up more volume in the shot—befitting not only his corpulence but also his implied wealth. The telephoto lens through which this frame is composed keeps only Andrews, viewed through the window, in clear focus, a visual strategy serving to narrow our sympathies toward a particular character. A cut to a closer shot of Andrews as she hungrily watches the man eat keeps her positioned within the frame of the window as melting snow trickles down the pane. Then follow a series of subjective shots, not quite from Victoria's literal point of view, of the man biting down on the dessert, and then a close shot of Victoria, still glimpsed through the window—she is about to faint. A return to the initial shot setup preserves the internal frame of the window, but now Andrews is not there. The downward glances of the passersby through the pane indicate their attention to someone below—Victoria, fallen to the ground. Her moment of intense hunger, as she stands, atomized, in the internal frame of the window, briefly gives way to some public acknowledgment of her hardship, in the form of the friendly people who help her stand upright after her fainting spell, even if they cannot much help her by way of food.

Some of the film's later, and more complicated, comic set pieces develop the initial play with internal frames utilized in these introductions of Toddy and Victoria. If the first scene with Toddy in the mirror suggests a private self slipping out of solitude to command a larger public frame—a public in *Victor/Victoria* implied by the lateral stretch and breadth of the anamorphic frame—the moment in the window with Andrews suggests the tentative nature of the survival of that self in the same world, a world marked by the economic severities of the Depression. Once Victoria begins her masquerade as Count Victor, however, she begins to enjoy acclaim and success. What she must strive to satiate in public, after her ruse as Count Victor proves economically successful, is not her hunger but rather her desire to continue to live illusively, both on the stage and off, as Victor. And as with the aforementioned sequences, her ability to do so will involve continually maneuvering around a mobile wide frame that gradually offers a wider range of positions and frames within which to live.

Midway through the film Edwards orchestrates a supreme variation on the "hiding under the bed" slapstick sequence—one that figures throughout his cinema, from the first *Pink Panther* to *A Fine Mess* (1986) to *Blind Date* (1987)—in which characters must intermittently hide themselves within and across rooms lavishly appointed with windows, beds, doorways, and curtains. The anamorphic frame offers a number of spots in which to hide, but also a number of ways the hidden figure might be caught, a comic tension especially key to *Victor/Victoria*. King Marchand and his burly but sensitive assistant, "Squash" Bernstein (played by former football player Alex Karras) are staying at a modishly decorated, art deco hotel across the way from the warmer, sparser, but equally lavish hotel in which Count Victor and Toddy, flush with success from Victoria's first performances, presently reside. Both hotels are replete with doorways, archways, and windows through which characters are temporarily framed and positioned as they seek to evade view. Marchand sneaks into Victoria's room to try to find evidence that his attraction to Count Victor is in fact attraction to a woman. In this sneaky sequence, Edwards positions the camera from within other rooms or spaces and toward Marchand, standing cautiously by the edges of walls or hiding within bathroom cabinets, framed by the visual pillars created by the vertical lines formed by the architecture of the room. Marchand is ostensibly slipping into this room to discover some evidence of what he takes to be the "reality" of Victor's biological identity—and he will indeed smugly and satisfactorily spy Victoria in the bath, from his hidden stakeout inside a bathroom cabinet, after she and Toddy return to the room while Marchand is there. But Victoria never actually discovers Marchand is here (after more machinations and in-frame hiding, he slips away from the gaze of both Toddy and Victoria and escapes the room); the "revelation" of her identity as Victoria to Marchand is in fact her own doing, later in the film. In this context, Marchand's ferreting away of himself within these internal frames works to hide *his* presence, and by extension his identity, which given his attraction to a potential drag queen takes on implications of homosexuality with which he is presently uncomfortable. Edwards underscores this aspect of the theme even further when "Squash" Bernstein, who will soon out himself as homosexual, follows Marchand into the room and finds himself similarly trapped within internal frames, mainly in the borders of the outdoor windows on the terrace where he hides from Victoria's and Toddy's view. Edwards uses this comic set piece not to reveal Victoria's "true" identity but to implicate the traditionally masculine characters in *Victor/Victoria*'s fluid conception of public and private identity, expressed throughout by a play with the placement of screen figures in shifting internal frames within the anamorphic image.

There are moments in *Victor/Victoria* when the internal security of the private frame gives way to something less veiled: musical performance with a choreography that takes up the entire lateral stretch and depth of the wide image, in the form of musical numbers performed variously by Victoria, Toddy, and Norma. In most of these numbers Edwards uses the wide image to establish a theatrical proscenium within the story world on which Victoria and the others perform, occasionally cutting away to shots of audience members' reactions that ratify Victoria's identity as Victor on the stage (see Lehman and Luhr, *Blake Edwards: Returning to the Scene* 51). The numbers performed by Toddy and Victoria in the Café de Luis, by contrast, take on an "in-the-round" form, with the mobile frame circling the café as the performers themselves circle about it, involving the audience in their performance. One exception to this general pattern in the staging of a musical number arrives near the end of the film, during Victoria's last performance as Victor before she gives up this public identity. In this sequence, the wide frame is now intimately involved with the onstage performance somewhat like it was in the opening sequence of *Darling Lili*. Edwards begins Count Victor's final performance on a close-up of a rose resting on a black piano, its petals the shot's only source of vivid, bright color. These flowers could be read as an oblique reference to the flowers in *Darling Lili* and, in this way, potentially implicating Edwards himself, and his relationship to Andrews, in the broader sexual play in this frame. From this rose the camera tracks to the hand of Victor and, in a gesture that associates the rose within Julie Andrews's eyeline, up to Count Victor's visage, which looks down at the piano. From here, Count Victor begins the song, turning to the audience as the frame continues its circling movement around him. This camera movement opens up Victor's performance not only to the audience within the film but also to Victoria's most important mentor: Toddy. As the frame circles to eventually reveal the audience in the music hall, for a moment Toddy, standing near the wings of the stage watching Victor's performance, is internally framed within the wide image, with a profile of Andrews and the red curtains to Robert Preston's right forming the borders of this fleeting internal frame. In earlier musical numbers, Edwards establishes the relationality between the onstage performance and the audience in the story world by cutting between shots of performers on the stage and viewers in the audience, as in the earlier sequence in which King Marchand watches, and falls in love with, Count Victor. This momentary presence of Toddy, viewing the performance from the wings, involves not cutting but rather a temporary internal frame revealed through the movement of the camera. This gesture implies the importance of Toddy to what Victoria has achieved in public, onstage throughout the film. That he is slightly out of focus in this particular shot makes the image all the more poignant, implying both his visibility and importance to Victoria as well as his relative invisibility—as a mentor and as a gay man—within the wider frame of the film's public world.

Toddy, through Edwards's writing of the character and through Preston's vibrant performance, is presented in *Victor/Victoria* as proudly out-of-the-closet. But the film, despite its obvious artifice, is also smartly aware of the limits bracketing Toddy's expression of his sexuality, which is acceptable in 1934 mainly in his private interactions with Victoria and in his public performances as a gay singer in the demimonde cafés of Paris. Toddy experiences the "internal frame" in the context of the larger anamorphic frame, in this sense, differently from other characters in the film. This fact makes the film's final musical number all the more exhilarating, as Robert Preston, as Toddy, finally *does* command the entirety of the anamorphic frame as he takes on Victoria's role as Victor after she relinquishes it in favor of a traditional heterosexual union. That the film should end by underscoring Toddy's hilarious, drag queen burlesque, rather than dwelling excessively on the romantic clinch between Andrews and Garland, demonstrates the nuance of Edwards's thoughtful meditation on sexuality and its various situations in both the private and public frames glimpsed in the film.

Displacing Dialogue in Edwards's Late Romantic Comedies

Although frequently associated with the slapstick antics of the *Pink Panther* films, Blake Edwards is perhaps best known for his hybrid interest in uniting physical slapstick with romantic comedy, work that begins with the generation-defining (and nonanamorphic) *Breakfast at Tiffany's* in the early 1960s. The films *"10," Skin Deep*, and *Switch*, discussed in the final stretch of this chapter, are all mature, yet still puckishly playful, reflections on the ways anamorphic frames can be imaginatively used in romantic comedy, a genre not frequently thought of in terms of visual technique.

In these late romantic comedies, we get the best evidence for what Adrian Martin describes in Edwards as "the thrilling possibility of actually thinking and instituting self-other relations in a radically different way: multiple-partner arrangements with the option of bisexuality" ("Blake Edwards' Sad Songs of Love"). Edwards discovers these "thrilling possibilities" not primarily through shared conversations between the couple—as in Stanley Cavell's conception of romantic screwball comedy (in his 1981 work *Pursuits of Happiness*), which emphasizes the importance, in classical films shot in 1.37:1 ratios, of shared conversation between members of couples as means for them to reach a state of reconciliation and reunion—but rather in, to cite a few more of Martin's words, "a sort of detached, almost purely formalistic inquiry into the mechanisms of narrative, character, and stylistic construction." This "formalistic inquiry" into the formation of the romantic couple involves simultaneously an inquiry in Edwards's late films into the ongoing creative possibilities of anamorphic cinema even after its standardization in the industry. We have already seen some of this at play in romantic comedies with Julie Andrews, *Darling Lili* and *Victor/Victoria*. The films *"10," Skin Deep*, and *Switch* are in this sense collective summations of how, for Edwards, romantic reconciliation is as much a matter of physical, bodily, and spatial—as well as verbal—negotiation.

"10"

In *"10,"* George Webber (Dudley Moore) is a successful composer and pianist. He is in a relationship Samantha "Sam" Taylor, a singer played by Julie Andrews. In early scenes, these characters (not married but nevertheless forming a relatively permanent couple undergoing an interpersonal crisis) bicker and quarrel, but it is apparent early in the film that no resolution will be discovered between them through the sharing of conversation. A reunion between George and Sam, for Edwards, can only occur through the physical stumbling and bumbling of George across the widescreen frame. George cannot explore alternative possibilities of life outside of his relationship with Sam through conversation with her, but must rather directly and physically experience the larger terrain of the anamorphic image. And only by doing so can George recognize that his relationship with Sam is the only form of coupledom designed for him.

A collaboration between Edwards and cinematographer Frank Stanley (who previously worked with Edwards on *The Carey Treatment*), *"10"* repeatedly underscores other ways of physically and sexually inhabiting the anamorphic frame and comically contrasts these with George's flailing. One of the most memorable bits of comic play in *"10"* arrives early. George, shortly after his forty-second birthday, becomes infatuated with a young, beautiful newlywed, Jennifer Miles (Bo Derek), whom he spies while driving. She is at that moment on her way to be married, dressed in virginal white and in this way fulfilling, as Lehman and Luhr have pointed out, George's ideal image of the perfect, untouched woman (*Blake Edwards* 247–248). Seeking spiritual solace, perhaps—but, more pointedly, wanting to know the name of the young bride he has seen in the church the previous day—George visits the reverend (Max Showalter) who presided over the marriage of Jennifer to her equally beautiful and equally young husband. The reverend invites George in for tea, which cues the presence of the very old, nearly blind, and very slow-moving Mrs. Kissell (Nedra Volz), whose efforts to bring the men a tray of tea result in her crashing into a fireplace on the other side of the frame. Lehman and Luhr have described this comical moment well:

> Suddenly, a little old lady's head appears in the extreme lower right hand corner of the frame. As if motivated by some unstoppable force, she precariously weaves her way across the frame at an excruciatingly slow pace. George and the minister [*sic*] pass her as they go to sit down. The camera pans to the left, following the two men who, after moving into the rear of the frame and sitting, continue their conversation. When we had almost forgotten her presence, once again the old lady enters frame right and totters across the frame with her tea tray. She passes in front of the men, who continue talking. We suddenly realize that the old lady is going to bypass the table and head straight into a large fireplace at the far left of the frame. She walks into it, slowly turns around, reorienting herself, and finally finds the table where she somehow manages to deposit the tea—but not without spilling it. After the minister tells her to return to the kitchen, she loudly breaks wind. (*Blake Edwards* 244–245)

What is especially funny, and ironic, about this moment is that despite her painfully slow movements and eventual crashing into the fireplace, this little old lady actually moves across the anamorphic frame more effectively than George Webber himself ever does in the film. While it is true, as Sam Wasson notes, that "Edwards uses the wide-angle lens, the anamorphic frame, and the imbalanced composition" to make the flatulent old woman's "journey appear to be longer and more arduous that it actually is" (211), this makes her eventual achievement of her journey no less comically impressive. Despite dropping the tray of tea, she does at least fulfill part of her goal—to move across the room. George attempts to have a conversation with the reverend in this scene, to discover a verbal solution to his problems. Mrs. Kissell, in comedic counterpoint, implicitly suggests George's verbal attempts of discovering a path toward resolution will eventually have to yield to ineluctably physical ones. But when he tries or is forced to move across the frame later in *"10,"* George will stumble and fall—down the hill on the side of his house, into the pool as he attempts to hang up a telephone, in whiplash after being stung on the nose by a bee while surreptitiously viewing

Jennifer at her wedding. In contrast to Dudley Moore's expertly and often vertically oriented and performed pratfalls (falling *down* the side of that hill)—the actor's brilliant expression of George Webber's physical incompetence—Mrs. Kissell is, in her own steady, windy way, and in collaboration with Edwards's manipulation of the wide image, something of a master of horizontal, lateral widescreen movement.

George's work as a composer also generates interesting ironies within Edwards's own practice as an orchestrator of widescreen compositions. We expect a piano player in a film to sit in front of the keys rather than moving boldly across an expansive space, and George Webber's seated position at his musical instrument in *"10"* parallels his similarly seated position as he looks at beautiful women, and at Jennifer specifically. Of course, the piano, as an instrument, can be filmed beautifully within this format, as Edwards demonstrates early in the film as George, serenading guests at his own surprise birthday party in the opening sequence, confidently tickles the keys of a piano spread across the frame. (Dudley Moore was himself a celebrated pianist.) But Edwards does not give us many more shots of Moore's piano prowess in which his musical ability alone commands the frame. Moore's later moments at the piano underscore the character's isolation within the wide image while seated at the instrument as he writes music, or his emotional solitude while playing love songs, during his impulsive sojourn to Mexico, as he plays while dreamily thinking of Jennifer. These seated moments of musical creativity and skill, which the young and inexperienced should admire, are in *"10"* melancholically contrasted to the physical beauty of youth. Such attractive bodies on display in the frames of *"10"* do not possess the creative talent, or years of experience, that George embodies, but they nevertheless command the frame of the film, and the focus of George's gaze, with their inherently breathtaking physical movements and gestures.

The day after his surprise birthday party, George visits his musical collaborator, Hugh (Robert Webber), a songwriter. The first shot in this sequence is a close-up of Hugh's gorgeous, and much younger, boyfriend, running shirtless on the beach, followed by a shot, positioned from a longer shot distance, of the young man running across the sand. This shot pans to follow the young man as he joins a group of equally attractive young people in conversation, a moment from which Edwards cuts to a shot of Moore at his piano, near a window, working on a new piece of music inside Hugh's beach house. He gazes out at the young people, but Edwards does not cut to a corresponding reaction shot of them. Instead, the shot composition places the reflection of Hugh's young boyfriend in the window next to Moore, a reflection that is itself internally framed in the image by the borders formed by the music stand holding the sheets on which George displays his musical notations (Figure 2.7). Edwards's composition not only suggests the distance between George's seated position and offscreen, youthful beauty but also suggests his frustrated desire is an energy source for artistic creativity. That the object of his gaze, directed offscreen, is in this moment not Bo Derek but rather Hugh's

FIGURE 2.7 *"10"* (Orion Pictures, 1979). Digital frame enlargement.

boyfriend also suggests the complicated sexuality George variously embodies and represses. Here, even homosexuality, not hitherto acknowledged by George as an aspect of his life in *"10,"* is implied as an alternative possibility in the composition of the anamorphic frame. This will be part of the physical and ambiguous terrain of his life that George must flail through before he can return to Sam.

Edwards also demonstrates in *"10"* his continued interest in a play with scale across the cuts connecting his separate widescreen shots and along the compositional depth of the anamorphic image. In such images, figures in the foreground and background of the frame become elements of salient contrast. Many of the most inventive and telling plays with scale across and within frames in *"10"* arrive in the film's beach sequences in Mexico, in which George, having tracked down Jennifer on her honeymoon, looks longingly and from afar at her. The representation of Bo Derek's body in relation to George in these frames is crucial. She is, throughout the final hour of the film, represented as generationally and psychologically *other* to George, a nearly complete contrast to his romantic fantasy of her as the virginal, ideal bride. She is, like Hugh's boyfriend but on a level that the film amplifies and dwells on more intensely, a physical fact that George, rather than merely gazing at, will actually have to grapple with in order to understand. He cannot just have a conversation with Sam, or anyone else, about it. This is an idea Edwards expresses through his orchestration of bodies in the wide frame in the beach sequences in Mexico.

Edwards's achievement in the beach sequences is to establish the contrasting presences of George and Jennifer through a play between the visual scale of contrasting physicality and movement in the anamorphic frame and through the placement of cuts that situate George's physical world as utterly apart from Jennifer's youthful realm. The very nature of the beach sends Moore spiraling across the frame: the sand is scalding, and as George burns his toes, Edwards's choice of a long apparent shot distance keeps in view a panorama of tourists who, unlike

George, are able to negotiate the terrain without stumbling. On this hot beach, George has no comfortable space wherein to sit and view Jennifer unobstructed. Edwards underscores in nearly every camera setup other physical presences surrounding George, implying other subjectivities, including a too-helpful waiter, other chattering hotel guests, and a hung-over woman named Mary Lewis (Dee Wallace) with whom George has had an unsuccessful dalliance the night before. Of course, Jennifer and her husband, David (Sam J. Jones), are two of these presences, both objects of fixation not only for George but also for Mary, who, in passing by the couple at one point, sighs longingly at David's suntanned, toned body. If the widescreen frame in cinema is potentially always a panorama enabling the viewer's free-roaming, desiring gaze, what George encounters here are more and more bodies incarnating more and more gazes, figures on a beach who are perpetually interrupting George's ability to immerse himself in the visual pleasure of Jennifer's body.

This ineluctable fact of a chaotically populated physical world for Edwards is the source of ongoing compositional fascination in his films, and it is the primary reason why, in his late romantic comedies, conversation between the couple is not enough to discover romantic reconciliation. A complex public sphere must be first invoked and then traveled through. George, of course, remains our central point of focus in the film's beach sequence. Nevertheless, the way in which Edwards frames and cuts those moments in which George is afforded a stretch of time to stare at Jennifer only reminds us how the expansive anamorphic frame exceeds the ability of any one subjectivity to master its contents. When David leaps into the ocean to relax on a surfboard, George has a rare opportunity to stare without interruption at Jennifer, who is suntanning on a towel several yards away from him. Yet in cutting from George's gaze to the object of his fascination, Edwards shows views of Bo Derek that could not possibly correspond to George's seated position. Shots of Derek, variously spreading suntan lotion over her arms and thighs, or turning over from her stomach to her back for an even tan, fill the anamorphic frame with Derek's physicality, the camera focusing in and fragmenting parts of her body in an intense and close way that suggests an impossible fulfillment of vision and access that remain unattainable for George. It is as if, in these particular wide frames, Bo Derek *becomes* the beach, incarnates in the form of a person the entirety of the physical terrain through which George has hitherto bumbled through.

But Edwards is here not simply giving the viewer an ideal position from which to gaze at Derek, in contrast to George's seated position of unfulfilled longing. He is showing how the desire to "master" an expansive and unpredictable frame with one idealized way of seeing—here, the wide frame's idealized composition of Bo Derek serving as a visual parallel and surrogate for George's frustrated desires to possess her—is always impossible in cinema, particularly in widescreen cinema. That these close views of Derek could not possibly correspond to any actually incarnated subjective point of view in *"10"* is further underscored by the fact that George will continue to daydream even as he apparently fantasizes about this "perfect" view of Derek on the beach, the widescreen frame momentarily filled with dreamy visions of Jennifer running toward the camera to embrace him. These fantasy shots, notably, include no one besides Moore and Derek in the frame, George's fantasy of mastery needing to do away with other subjectivities—including, ultimately, Jennifer's own, even though she is physically present in his illusions. But George can only find reconciliation with Sam through his ultimate physical movement through the reality of the landscape around him, rather than in illusory images that can only defer the ultimate necessity of moving through the reality of these frames.

In this way, *"10"* becomes Edwards's own form of therapeutic anamorphic cinema, a film in which the director uses the wide frame to diagnose, for comic effect, one middle-aged man's tortured subjectivity as he journeys through a world of young, nubile bodies. The capacity of the Panavision frame to include the physical presence of other implied subjectivities, and to enable a panoply of ways of looking at and navigating desire, becomes the very terrain George must negotiate before he realizes that it is Sam, and not Jennifer, with whom he has a rich future. Edwards's widescreen comedy of romantic reunion in *"10"* acknowledges the physical components that must be addressed in romantic life, the realities of sexual desire and the existence of alternative ways of physically inhabiting this big, wide world.

Skin Deep

Skin Deep, appearing in cinemas a decade after *"10,"* forms something of a spiritual counterpart to the earlier movie (it was originally written as the literal sequel). John Ritter plays Zach Hutton, a once-successful but now stagnating novelist—and, like George Webber in *"10,"* a talented piano player. Zach, when the film begins, is embroiled in a series of love trysts despite being married to Alexandra "Alex" Hutton (Alyson Reed), a newscaster. In the opening scene, one of Zach's many lovers, Angie (Denise Crosby), happens upon Zach in bed with another woman, his hairdresser, in the home he shares with Alex. On a narrative level, the film eventually works, as comedies of remarriages do, to bring Zach and Alex back together again. However, on the level of the film's anamorphic style (created by Edwards with cinematographer Isidore Mankofsky, in his only collaboration with the director), the film reminds us that this reunion cannot simply happen through Zach and Alex engaging in conversation. Indeed, the conversations shared between them are even more perfunctory than the ones between Dudley Moore and Julie Andrews in *"10."* Edwards frames the problem facing Zach in *Skin Deep* as once again one that cannot find resolution through words. A friend of Zach's, Sparky (Peter Donat), suggests that Zach needs to focus on putting more of his witty dialogue into his novels rather than using his powers of wordsmithery to throw verbal barbs at cocktail parties. Zach, who remarks during the same party that he is there just to see if any "narrative possibilities" emerge, seems aware that his problems in life are connected to his problems in writing. For Edwards, Zach's problems are precisely those involving the presence of the body in the anamorphic frame. If the play with scale across anamorphic frames in *"10"* involves a presentation of Bo Derek's body as ultimately impossible terrain for George, in *Skin Deep* the body that Zach is struggling to possess is his own. Unlike George, the middle-aged Zach has no trouble at all bedding beautiful young women. He speaks about his desire for monogamy, but ultimately Zach's behavior expresses none of the decorum ultimately preventing George from having sex with Jennifer after she shows an interest in him near the end of *"10." Skin Deep* very clearly posits Zach's problem as a problem of the body—*his* body. His desire to have affairs with many beautiful women, despite his professed and simultaneous longing for a stable romantic relationship, is presented by the film as a physical desire the anamorphic image cannot contain, and a desire, despite its expansiveness, it cannot satisfy.

This is evident in the aforementioned scene in which Angie happens upon Zach in bed with his hairdresser, a conflict that Alex herself disrupts when she returns to the house. As Zach is being pinned to the bed in a headlock by his vexed former lover, Edwards frames John Ritter's prostrate body so that it occupies a significant portion of the lateral stretch of the anamorphic frame in the middle and bottom right portion of the image. But this framing also underscores how Ritter's very body cannot possibly be fully contained by it. Observing the rule dictated by the Motion Picture Association of America (MPAA) for the presentation of male genitalia in his framing of the shot, Edwards elides any view of Ritter's nether regions while underscoring the way in which the actor's visibly naked body nevertheless still engulfs a significant stretch of this image. But Edwards's framing here is not simply, politely obeying the MPAA's stringent double standard regarding the presentation of frontal nudity in films. Later, in the glowing-condoms scene of *Skin Deep*, in which the film displays for us, in darkness, two glowing, erect penises swashbuckling at one another in a riff on the lightsaber duels in *Star Wars* (1977), Edwards will poke fun at precisely this form of repressive censorship. But in this frame from the opening sequence of *Skin Deep*, he nevertheless manages to show quite a bit of John Ritter's body while also underscoring the fact that this frame cannot contain all of him, nor all of what that body desires. Zach will need to conquer this problem posed by the situation of his body in the anamorphic frame precisely in order to realize the more contained and perhaps inherently repressive form of living signified by marital monogamy.

Later, Zach reflects on events we see elsewhere in the film as he talks to his therapist, Dr. Westford (Michael Kidd), about his struggles to commit. Even in the therapist's office, though, in which Zach is supposed to be in a seated position and focused on conversation that might help him conquer his problems, he is faced by the recurring problem of his body and his inability to understand his desires through talking, even in the ostensible expanse offered by the wide image. In the first scene with the therapist, Zach initially is seated, but he eventually gets up and breaks out into a satire of a musical number, meant to show to his therapist the inability of psychoanalysis to provide a resolution for his obsessions. His comical jig motivates a leftward pan of the camera, as Ritter dances around the therapist's chair, and an ensuing rightward pan, in which he exits the room like a dancer leaving the stage after a showstopping number. Zach's dancing body on the one hand indicates his stubborn and comical unwillingness to repress his desires, through a sudden irruption of outlandish musical performance. In doing so, he also provokes the anamorphic frame, which dutifully follows him leftward and rightward, to acknowledge its own inability to understand the physical complexity of its central character in a stationary wide view.

Similar movements, of both Zach and the laterally mobile wide frame, also occur in the bar Zach frequents, an alcoholic's alternative to therapy. He frequently sits at the bar with Barney, the bartender (Vincent Gardenia), and discusses his problems, conversations Edwards frames in shot–reaction shot setups that saliently activate only the middle portion of the anamorphic frame. However, Zach's compulsive return to the bar's piano—placed on the right side of the mise-en-scène, away from the main bar—underscores the way in which his body also becomes a problem for him even in these therapeutic scenes. He moves compulsively from a position of imbibement, on the left side of the bar, to a position of musicality, on the right. Zach's fidgety movements to the piano—a metonym for his refusal to focus on his writing—also serve to open up the anamorphic frame for the arrival of additional women. This is evident when Zach goes to play the piano as a female bodybuilder, Lonnie (played by Raye Hollitt, one of the American Gladiators), arrives in the bar. The arrival of Lonnie, in terms of the film's anamorphic logic, has been enabled by Zach's movement toward the piano. Prior to his walk to the piano and the camera's ensuing reframing of the scene, we have been generally unaware of the entrance to the bar and the possibility that another beautiful woman might walk through it. Lonnie will become attracted to Zach's piano playing, and her movement into the bar, and eventually toward Zach, continues to activate the frame's lateral pans. The problem and presence of Zach's body, and the various manifestations of his presence (as psychoanalytic patient; as song-and-dance man; as a flirty drinker in a bar; and as piano player) not only are caused by his unruly interest in women but also work to perpetually generate social spaces in which he will meet more and more of them.

The film repeatedly calls attention to how even these generously endowed anamorphic frames neither contain nor satisfy Zach. After an evening's tryst with the bodybuilder (her bedroom is adjacent to a gym she owns), Zach awakens to discover she is giving an aerobics class to a room full of women. Zach, in his underwear, swallows his embarrassment after he accidentally locks himself outside the bedroom door. Joining the women, clad in their aerobics leotards, Zach becomes, alongside them, as if a member of a chorus line (Figure 2.8). Once again, Zach's body has found itself in a frame in which his mere presence and

FIGURE 2.8 *Skin Deep* (Twentieth Century Fox, 1989). Digital frame enlargement.

movements enable the sudden appearance of more desirable women. Although the lateral stretch of the frame here—full of precisely that which Zach desires—comically accommodates and "contains" Zach in his "workout," the scene of him performing aerobics in his underwear is of course absurd, demonstrating the comical situations he flails into throughout the movie.

Ultimately, Zach solves his problems with women through another meeting with his psychoanalyst, who advises him to stop drinking. Zach is seen drinking in nearly every scene of social interaction in the film, underscoring the extent to which *Skin Deep*, like Edwards's earlier *Days of Wine and Roses* and *Blind Date*, is about alcoholism. His analyst also advises him to spend a lot of time thinking about his future while sober. These words of wisdom, indeed of conversation, form the basis for Zach's eventual reunion with his wife, Alex. But, again, and despite the plethora of witty dialogue in *Skin Deep* and despite the fact that Zach seems to find a path to a reunion with Alex through the discussion with his therapist, Edwards de-emphasizes the importance of conversation between Zach and Alex in the final stretch of the film. Zach has his "discovery" that he must quit alcohol while he is sitting alone on the beach, as he gazes out on the ocean while reflecting upon his therapist's words. A tidal wave is coming in, but Zach is focused on his "eureka!" moment, when he realizes he must stop drinking, commit once and for all to Alex, or else be left alone to face the absurdity of life in solitude. But these words of commitment matter less to Edwards than does the physical and visual comportment of Zach as he discovers them, with a bursting tidal wave of water exploding into the room of a beach house while he exclaims his belated discovery of his need to recommit to Alex and to his writing. (Or, as he puts it: "There is a God, and he's a gag writer!" God's jokes transcend mere human talk.)

Because the reunion of the couple takes place within the context of a wider anamorphic image that must go beyond a narrower, more intimate scene of conversation, *Skin Deep*'s final sequence in Barney's bar does not just feature Alex and Zach reconciling. The scene is a celebration of Zach's publication of his novel; his giving up drink and getting back to the written word are the keys to his rediscovery of the pleasures of monogamy. Ironically, even though Zach seems to have outwardly solved his romantic problems, Edwards emphasizes even at the end of the film the way in which Zach's body still cannot be contained by a single anamorphic frame. In this penultimate sequence of the film, Zach has gathered together his friends, as well as his many previous lovers, to toast his literary achievement. John Ritter moves around the room, and as he does so the camera follows him. Zach's body, although now apparently disciplined through his publication of new writing, is nevertheless a presence motivating the anamorphic frame's movement. He still desires the pleasures of social interactions and flirtation, as a brief exchange at the piano with a beautiful young literary scholar Rebecca (Jean Marie McKee) suggests. On the level of the narrative, Zach's body problem has been solved. He is committed to his writing, and to one woman. He (apparently) resists Rebecca's flirtations. But on the level of images, Zach's body will still not be contained by any single position, and will continue to encounter such problems as he moves through these frames. He will still, in other words, continue to move toward that damn piano, which seems to always enable another erotic, musical moment of flirtation, even if Zach is now in a better position by film's end to resist its fulfillment.

The final sequence of the film, which glimpses Zach and Alex in bed at home at some point after the party, gives audiences, on the surface, a quite conventional happy ending, a reunion of the heterosexual couple committed to an apparently monogamous future. Nevertheless, after the couple turn the lights out, Zach puts on a variation on the glowing condom from the earlier scene in the film. Now his erect penis, lit up in what is otherwise the engulfment of complete darkness in the anamorphic frame by what Zach calls a "patriotic" red, white, and blue glowing condom, is the only thing we see. The film has ended by suggesting that the problem of Zach's body—here, literally, the problem of his ongoingly erect penis—has been successfully contained by the expansive anamorphic frame. It no longer darts about the space of the room as it did during the earlier, comical scene in which Zach duels with another glowing, condom-sheathed penis when another man discovers Zach in a hotel room with the man's girlfriend. By referring in this way to an earlier scene in which such containment only led to animated conflict across the lateral stretch of the frame, the film now implies, despite the ostensible "containment" of Zach's desires in this final frame, the ongoing problem that will be posed by the reality of Zach's innate desires. And, of course, this glowing condom, full as it is with a male member engorged by blood (precisely what we are *not* supposed to be seeing in an R-rated Hollywood movie, according to the MPAA's repressive strictures), is a sign of Edwards's own ongoingly physical and, as much as is possible in the inherently censorial Hollywood system in which he works, unrepressed play with the wide image.

Switch

In contrast to *Skin Deep*, in the final romantic comedy of Edwards's career, *Switch* (which reunites Edwards with frequent collaborator, cinematographer Dick Bush), the problem is not so much the body's resistance to containment within the wide frame but rather the containment of one type of person in a biological body altogether unfamiliar and alien. The wide frame, rather than serving as a border across which the excesses of the male body always spill as they did in *Skin Deep*, becomes a more centripetally oriented proscenium through which we view newly unsettling embodiment. In *Switch*, rather than "spilling out" across the borders of the Panavision image, social pressures and situations constantly form sites of pressure on a body in which the soul of a man has become reincarnated in a woman, with the "soul" perhaps changing the contours and nature of its identity as a result. The film relies heavily for its comedy on the brilliant performance of Ellen Barkin, and on her estimable and frequently hilarious ability to physically incarnate the soul of a misogynistic man in the body of a woman, a body of which this man of course has no understanding. The film is replete with funny moments in which Barkin, conveying a sense of how her male character, Steve, becomes acclimated to his soul's reincarnation in a woman's body, after he is murdered by three women seeking revenge on him for his sexism, contorts and positions and gropes and awkwardly moves her own body in order to convey to the viewer the idea of a masculine spirit inhabiting the inwardly and outwardly unfamiliar terrain of female biology. Edwards's wide frame in *Switch*, in contrast to *Skin Deep*'s emphasis on the way the male body exceeds and spills over the borders of the frame, is repeatedly, masterfully positioned to ideally depict Barkin's comic characterization of a man quite unable to adapt to conventions of normative feminine bodily presentation. In one meeting with her new boss at an advertising firm, for example, Amanda—as Steve has taken to calling himself after being reincarnated—sits variously in ways that demonstrate a lack of self-awareness about how to normatively cross her legs in "polite" company, and in each of these shots Edwards's camera is perfectly situated to capture the ensuing moment.

But *Switch*'s use of the wide frame goes beyond situating it merely as the perfect proscenium for Barkin's brilliant comic performance. The film subtly patterns a series of frames in which the "split" nature of the male-coded and female-coded personalities vie for dominance within Steve/Amanda, a body-mind bifurcation metaphorically paralleled by a series of visual bifurcations in the anamorphic frame. These bifurcated images trouble the idea that the wide frame of *Switch* is only being used as a display case for Barkin's performative skill and intensifies our sense that there is an artist, Edwards, deepening the implications of the film's comic premise through his anamorphic stylization. The bifurcated anamorphic frames of *Switch* are somewhat different from the approach to the plethora of internal frames in *Victor/Victoria*. There, relatively stable if private sexual identities become intimately legible in the internal frames of mirrors, windows, and doors and are socially asserted through a variety of public prosceniums (the stage performance at the end of the film, for example, in which Robert Preston, freed from his relatively marginal isolation in the anamorphic image, commands the entire stretch of the public frame). By contrast, *Switch* does not suggest that Steve-turned-Amanda is of necessity "hiding" some kind of private subjectivity from the public frame. Instead, she is working through the implications of her sudden switch in the context of what is now, in light of the reversal, freshly complex social terrain. Within this world Edwards finds visual bifurcations—sometimes across the lateral stretch of the image, sometimes along its depth—that intensify the split in Amanda even as they encourage us to take the film's meaning beyond the binary contours of its premise.

The opening of the film, in which the male incarnation of Steve (Perry King) cavorts with three women, is notably absent these salient bifurcations of the frame, as if to suggest the extent to which Steve himself sees his widescreen world as a stage in which he might find the presumed pleasures owed to him as a heterosexual man—here, sex with three women in a hot tub. These women, however, have turned the tables on him. What was taken by him as his usual proscenium for hedonistic sex is actually a little bit of theater they have put together to lure him into a trap and kill him in revenge. (The film does not reveal the precise crimes Steve is accused of, but King's performance makes clear his smarmy misogyny.) The subsequent presentation of the purgatory between heaven and hell, in which Steve finds himself after his murder, begins the film's patterned use of bifurcated anamorphic frames. Edwards's presentation of purgatory begins with the lone, naked, wet figure of Steve, positioned in the middle of an extreme long shot and hopping on an unseen surface, surrounded by utter darkness. The image seems to be an answer to the question formed by *Skin Deep*'s problem of the body that cannot be contained in the frame: here, in the purgatory of *Switch*, the body can indeed be completely contained by the frame but only in a spiritual world apart from ours, waiting to meet its judgment by unseen gods and goddesses. There follows a cut to a medium shot of Steve against the darkness, looking around in anguish. "Where am I?" he asks, a reasonable question. At this, Edwards cuts to a bifurcated frame, with a shaft of ethereal blue and white light extending from the bottom of the lower left side of the image and up to the right corner, with Steve again positioned diminutively, across from this heavenly light, on the right. A male and a female voice alternately explain the nature of Steve's predicament: he is dead but can only be admitted to heaven if, upon being reincarnated and returned to earth, he can prove that at least one woman actually *liked* the man he was. As this conversation unfolds, Edwards varies the visual presentation of this bifurcation, at one point positioning the camera directly to the side of the spirits and Steve. In his patterns of cutting these images—which alternate between close shots of the beaming light and Steve—Edwards, despite the supernatural nature of the subject matter, closely follows the rules of continuity editing in the shot–reaction shot patterns, treating the spirits themselves as if they were forms of embodiment rather than spirits that might transcend even the flexible laws of cinema. This complements the dialogue shared between Steve and the gods, in which they reveal the limited nature of their omniscience (they are gods and thus should they not already know if there is *one* woman out there who likes Steve?) as well as their own bifurcated identities. In alternating between the voice of a man and a woman, this ethereal light in *Switch* "embodies," in its status as both male god and female goddess, the same predicament that will beset Steve when he becomes Amanda in a subsequent scene.

These voices of heaven have not insisted that Steve's reincarnation be as a woman, however. That sneaky trick is an invention of The Devil (Bruce Payne) who, unlike the ethereal light through which the spirits of heaven manifest themselves in *Switch*, is embodied in human form and visits Steve on earth, in the scene immediately following the purgatory sequence, to trigger Steve's biological "switch" to womanhood, to Amanda. The Devil perhaps believes that Steve's job of finding a woman who likes him will become even more difficult if he takes on the very body of the entire gender he has abused as he attempts to do so. Here Ellen Barkin makes her entrance into *Switch* as Edwards continues his patterned use of bifurcated frames. In this sequence, the bifurcation is between Steve's new incarnation as Amanda, on the one hand, and a figure of the Law, on the other. The security guard of the building in which Steve lives has been called after a report of a woman's screams emanating from his apartment. The screams resound after Steve discovers the female nature of his reincarnated body (which he will christen, subsequently, as Amanda). The security guard, once inside the apartment, is positioned throughout the first part of the sequence on the right side of the frame, while Amanda is variously figured as darting across it in a panic, before she stops to contemplate the situation, settling in a position on the left side of the frame, leaning against the wall. Edwards pointedly includes a mirror in the middle of this bifurcated image, situating this object, so frequently in films a metaphor for identity crises, as the midway point. The frame, from this position and in the same shot, now begins to track again across the apartment in a semicircle as it follows Amanda to another mirror, with the security guard looming rather haplessly in the background. Here, she sees her reflection, with the guard now reflected in it, another bifurcated anamorphic image now marking a vertical division, in the form of a line in the mirror, between the body of Amanda on the left and her mental reflection, signified in the mirror, on her situation on the right. Any grappling with her freshly discovered identity as a woman, this image efficiently suggests, will involve working through the norms and expectations—the Law—for her behavior, stringent rules she will flout as she continues to behave as the brashly confident Steve, but now in a body in which strident forthrightness is not always socially accepted.

Amanda more or less reconciles herself to her new embodiment and ventures out into the social world Steve once called his own—a bar he frequented and the

New York City ad agency that is his workplace. Her arrival at the ad agency results in one of the most complex shots in *Switch*, a three-minute-and-eleven-second-long take that alternates between medium-close shots and long-shot distances of the actors, as the camera variously tracks laterally and across the depth of the image, and as the actors themselves move across it and along its depth planes. All movements in this shot pivot around a bifurcation that Amanda, in turn, confidently complicates through her movement across and within the frame. Amanda visits the ad agency and reveals to Steve's coworkers that he has disappeared, and that she is looking after his apartment. Here Edwards creates compositions that momentarily cleave Amanda from the sites of unquestioned corporate privilege Steve previously occupied—his office—a complex bifurcation that separates Amanda not only laterally, in the Panavision image, from Steve's meek secretary (Catherine Keener) but also in terms of the depth of the shot. Amanda is separated from the former office in which Steve worked and in which his coworker and friend, Walter (Jimmy Smits), in the background of the frame, is now rummaging, for some kind of note from this office's former inhabitant that perhaps will help explain Steve's sudden disappearance. Edwards keeps the action in a single long take as Amanda makes her way slowly into her old office, stepping, with a pronounced wobble indicating Steve's ongoing attempt to learn how to walk in Amanda's high-heeled shoes, through the door into the private office that forms the midpoint of the bifurcated image. Bobbing into the office, Barkin exchanges positions with Smits, who now stands outside the door in the position previously occupied by Barkin as she steps inside and rummages through Steve's things. She plants and then "discovers" a fake note she has written explaining Steve's disappearance (he has fled to exotic lands to paint like Gaugin, an interesting notion given the intensely visual way Edwards explores the question of transgendered embodiment at play in *Switch*). While Smits reads the note, Barkin stands in the very midpoint of the anamorphic image, playing with the golf putter. Her position at this midway point in the frame suggests not so much that Amanda herself is bifurcated but rather that she is now able to cross these physical barriers that signify corporate and gendered hierarchy (Figure 2.9). When Steve's secretary begins crying at this news, Amanda leaves the office to comfort her. Barkin and Keener are now slightly closer to the camera as the frame has tracked in toward them, capturing their brief conversation in a medium shot. Keener's secretary is, actually, intensely happy that Steve has disappeared (she always cries when she is happy, as she informs Amanda and Walter). She leaves the office, the tracking camera following her to the door.

At the secretary's exit, Steve's boss, Arnold Freidkin (Tony Roberts), arrives through the doorway, all in the same long take. Edwards tracks back to the former position of the camera, with Amanda now inside the office as Walter and Freidkin discuss Steve's sudden disappearance. The two men stand on either side of the doorway, while Amanda, eventually, ventures back inside the office to play

FIGURE 2.9 *Switch* (Warner Bros., 1991). Digital frame enlargement.

more golf. Walter, as played by Smits, is a traditionally masculine man whose own hints of misogyny are more subtle than Steve's. He is a likable, somewhat dopey fellow who is nevertheless capable of sensitivity. And the film's positioning of him at various points in the frame previously occupied by Barkin (pirouettes of exchanged staging that occur throughout this shot's orchestration of mise-en-scène) prefigures the way he will become romantically entangled with and eventually committed to Amanda, even after he becomes convinced, at the end of the film, that it is indeed Steve's masculine spirit inhabiting her body.

The film will continue to play with a variety of bifurcations, although with the increasing sense that Amanda has learned how to occupy her social terrain with puckish unruliness. In one shot, for example, she discovers how to relax in her office (something which Steve, who early in the film never appears to actually be doing any work, never struggled with), stretching out her legs on her desk. The right side of the frame, occupied at various points by her subservient secretary, is connoted as "feminine," but the lateral stretch of Barkin's body across the image demolishes this division, creating a new sense of how her body might work in this space. Interestingly, bifurcation emerges again when Amanda begins a romance with a client of the ad agency, Sheila Faxton (Lorraine Bracco), who takes an interest in Amanda. In one scene in a swank nightclub, any sense of binary division is dissolved by the frame's presentation of happy dancing couples and indeed by a wall replete with images of classical Hollywood stars, many of whose complex and ambiguous performances have served for decades as imaginative inspiration for gay communities. Amanda and Sheila eventually butt heads, however, and are presented by Edwards in various bifurcated frames when it becomes apparent that both have headstrong, "masculine" personalities. Of course, this being a Blake Edwards film, their verbal parries ultimately give way to a comic slapstick fistfight involving surrounding dancers at the nightclub. Nevertheless, these "divided" images suggest not one more inscription of the primarily heterosexual conflicts signified earlier in *Switch* but rather fresh conflicts between two women with determined spirits, a new form of bifurcation.

The various conflicts described in the aforementioned frames are largely private in nature, or are at least limited to the relatively private corporate world in which Steve once worked and which Amanda must now navigate. The Law, figured early in the film by the presence of the security guard watching over Amanda upon her discovery of her new embodiment, emerges again near the end of *Switch*. Steve's dead body is discovered in the Hudson River, and Amanda is the primary suspect. By this point, Amanda and Walter have become romantically involved, although the ensuing pregnancy is unplanned. On trial for Steve's murder, Amanda is found to be insane and is relegated to a mental institution, where she gives birth to her child. The film here provocatively approaches a number of questions, including sexual consent (the film presents Walter's tryst with Amanda ambiguously, very possibly as a rape) and the right to choose. At this point in the film, the binary conflicts inscribed in the various bifurcated frames throughout the film have largely been overcome by Amanda, but at the cost of her social position: found insane, she is sent to a mental institution. At the end of the film, Edwards largely abandons bifurcated compositions, now relying on emotionally moving close-ups of Barkin as she explains to Walter her decision to keep the baby and her desire to marry him. That both Walter and Amanda are aware that Steve's masculine spirit is "haunting" Amanda's body charges this ostensibly heterosexual union with impish mischievousness. These close-ups of Barkin as she declares her love for both her baby and Walter abstract her from the surrounding space in the frame, which earlier served as the proscenium for most of the film's physical comedy and various bifurcations of the wide image. These moving close-ups also suggest Amanda's own psychological and personal resolution of the binary conflicts besetting her throughout the film. *Switch* pointedly makes clear, however, in its abstraction of Barkin from the surrounding space of the mental institution to which she is now doomed to live the rest of her life, that what one individual resolves may nevertheless be ongoingly contested by the various strictures of her society.

Amanda, upon giving birth to her baby daughter, dies. She has, apparently by this point in the film, fulfilled the charge given her in purgatory, that is, to find a female who likes her. Perhaps her daughter is the one female who finally likes Steve. Or, perhaps, Amanda is herself a substantively distinctive enough identity at the end of *Switch* to be said to be the very woman who has discovered a love for the man she used to be. But this source of fresh affection, of course, is also partially the result of the love of Walter (perhaps revealing a love for Steve, now that his spirit is safely enclosed in the body of an outwardly attractive woman, that he was previously unable to acknowledge in Steve's prior male incarnation). Perhaps even Sylvia Faxton's earlier affection for Amanda had something to do with the decision of the spirits in heaven to allow Steve's spirit to enter. In any event, the birth of the daughter and Amanda's recognition of the

FIGURE 2.10 *Switch* (Warner Bros., 1991). Digital frame enlargement.

infant's love for her cement the deal. In the final shots of the film, Walter and his daughter visit Amanda's gravestone. A close-up of the marker confirms that Amanda Brooks was "a great guy and a very special woman; may they rest in peace." Edwards has done more here than presciently anticipate the anxiety over pronouns in American culture that would emerge two decades after *Switch* was made, in the years following Edwards's own death in 2010. In this shot and the subsequent shots of the city skyline and the blue skies of heaven, we hear Barkin's voice still mulling over whether she would like to take on the form of a man or a woman in her afterlife in heaven. In the final shot of the film this voice-over resounds over the New York City skyline, the Twin Towers, now-lost emblems that in this context architecturally incarnate the kind of balance Amanda/Steve might be searching for, looming on the left side of the image (Figure 2.10). Her voice, heard on the soundtrack accompanying this blue sky dotted by clouds, makes clear that she decides, at least for now, not to decide. The film itself, having playfully engaged a series of complex bifurcated images involving questions of gender identity in American society, settles on this vision of indecision, of in-betweenness, as a kind of heavenly ideal. For Edwards, this ideal embraces, openly but also in an existentially ambiguous way, both masculine and feminine values. With that, *Switch*, and Edwards's charming mastery of anamorphic cinema, is brought to a close.

3

Robert Altman (1925–2006)

Diffusive Widescreen

In *The Long Goodbye*, we didn't show people what to look at. It was [*draws a Panavision frame across the air*], and you have to look among all these things, decide what's important; but you won't necessarily know it at the time.
—Robert Altman (qtd. in Smith and Jameson 172)

Films discussed in this chapter: *M.A.S.H.* (1970); *Brewster McCloud* (1970); *McCabe & Mrs. Miller* (1971); *The Long Goodbye* (1973); *California Split* (1974); *Nashville* (1975); *3 Women* (1977); *Short Cuts* (1993); *The Company* (2003).

Beyond Figuration

Robert Altman's first major success in Hollywood, the antiauthoritarian war comedy *M.A.S.H.*, resists conventional uses of widescreen. Many war movies shot in the anamorphic format tether audience identification to a handful of characters—usually depicted as being on "our" side. *M.A.S.H.*, by contrast, initiates Altman's career-long insistence on refusing the viewer straightforward identificatory entry into his diversely populated widescreen worlds. Filmed in anamorphic Panavision by cinematographer Harold E. Stine, like Altman a veteran of filmmaking for television prior to his arrival in Hollywood, *M.A.S.H.*'s representation of screen space is a tricky and tantalizing combination of lateral complexity and cinematographic compression. Robert Kolker describes Altman's style as encompassing a "wide, shallow space" (333) through the combination of anamorphic frame and frequent employment of telephoto lenses. This technological pairing opens up screen space laterally just as it compresses foreground, middle ground, and background into a squelch of cacophony. From out of this jigsaw arrangement arrives divergent positions and points of view and an eschewal of workaday coherence. Altman welcomes the democratic participation that André Bazin saw as one of the promises of CinemaScope in the 1950s, a technology that promised to capture a wider view of an already inherently ambiguous social reality. The director's widescreen work posits the possibility, however, that the very nature of "democratic participation" may itself be ambiguous. A viewing sensibility not open to playful irony—to the idea that the perspective one has "discovered" in Altman's widescreen world in any given shot might be turned on its head in the very next one—will be stymied by these films. If Jean Negulesco's films playfully defer meaning and fulfillment via the aesthetics of the wide frame, they are nevertheless held together, retrospectively, by the romantic clinch that ends most of his widescreen films. For Blake Edwards, the individual character is thrown into the absurd swirl of chaos that defines society in his cinema, even as the character remains a relatively stable focus point. Altman's wide frame, by contrast, suggests that the world as it arrives to us is already diffuse and polyphonic, rendering any gesture of deferral redundant. Altman does not presume that the world, even when glimpsed through the ostensible expanse of the Panavision frame, will bend to the descriptive or interpretive will of a beholder. His anamorphic cinema, which plays in the space between modernism and postmodernism, depth and surface, asserts the possibility that our creative participation in perceiving these frames may lead to no meaning at all.

Meaning in *M.A.S.H.*, indeed, seems entirely negative, a reaction to an external force—the perceived abuses of authority—rather than emerging from internal, generative energies. The wider anamorphic screen, a vast space across which characters might act, is in Altman not often a site for viable effect. Even particular character traits are eventually situated in the wide frame as part of a general social milieu, rather than as reflective exclusively of individual psychology. The caustic and antiauthoritarian dispositions of two of *M.A.S.H.*'s central characters, the military surgeons Hawkeye Pierce (Donald Sutherland) and Trapper John McIntyre (Elliott Gould), are not presented by the film as unique personality traits but rather as general attitudes shared by nearly all the characters in the ensemble. In selecting the 2.35:1 aspect ratio for *M.A.S.H.*, Altman uses a frame shape familiar to war spectacle. In doing so, he has thoroughly ironized the conventional relationship between subject matter and this cinematographic format. Rather than using the anamorphic frame to immerse viewers in a restaged cinematic war, Altman's images in *M.A.S.H.* generate pictorial absurdism that objectifies the human figure in the frame and diffuses any concrete sense of literal-minded, psychological representation. What results is constant visual transfiguration and perspectival contingency.

One of the film's signature images is a laterally arranged shot composition, depicting several characters seated at a long dinner table in sacrilegious evocation of Leonardo da Vinci's *The Last Supper* (1495–1498). Although outwardly parodic, this frame is situated in a grave narrative situation. The M.A.S.H. unit's dentist, Painless Pole (John Schuck), fearing a lost ability to perform sexually with women, outrageously declares intentions to commit suicide. His fellow medics craft an elaborate tableau to bring him back to a desire to live. Their staging makes astute use of a wide proscenium, positioning Painless as a Christlike figure who believes he is about to take a lethal pill as he lies in a casket, waiting for impending death. Altman's cinema, notable typically for its combined, staccato movements of figure, lens, and camera, here presents the actors in a tableau in which they gesture minimally, holding narrative progression in momentary suspension. The characters themselves have apparently gone to a great deal of theatrical labor to create this mise-en-scène, but while the film shows us the causes motivating this exercise—the revelation of Painless's worry about his possible homosexuality and the apparent existential threat this poses to his ongoing existence in the army—we do not in fact see any of the characters create this painterly moment as they might if they were stagehands or prop masters on the set that the film *M.A.S.H.* itself is. This evocation of *The Last Supper* does not simply operate in excess of the narrative information it is presenting but also questions, through the very salience of its painterly and theatrical inflection of the wide image, the reality status of the fiction itself.

While Altman is fond of discussing the art of painting as an influence on his filmmaking technique, he rarely explicitly cites works of art in his films. He also rarely emulates the look of representational painting, eschewing classical arrangement of mise-en-scène and traditional lighting. In certain of his anamorphic films the characters are themselves artists—a painter in *3 Women*, dancers in *The Company*, musicians and singers in *Nashville* and *A Prairie Home Companion* (2006). In those films, Altman's widescreen style will work not to replicate the characters' artistic practices but rather to portray a complex social milieu in the frame, a world in which his artist characters' work circulates and is received. Altman's style as a filmmaker eschews the stable onscreen representation required to create these explicit citations of paintings in cinema; this antiauthoritarian "reproduction" of *The Last Supper* in *M.A.S.H.* is unusual in that regard. His work in Panavision, the widescreen technology Altman uses most frequently throughout his career, tends more often toward images Joe McElhaney, in a discussion of *3 Women*, characterizes as "soft and diffuse" (147). Figures momentarily glimpsed in Altman's pictures very quickly slip into other forms as Altman's movement of the frame—through either a mobile camera or an optical movement, such as the use of the zoom lens—takes us quickly beyond a given moment's arrangement. This aesthetic diffusion situates the director's use of the wide Panavision frame as particularly porous and always in a state of undulating transformation (see Hamish Ford, "The Porous Frame").

Even this relatively legible citation of *The Last Supper* in *M.A.S.H.* ultimately avoids representational stability. The shot begins not with *The Last Supper* tableau itself but rather with a conversation between the unit's anxiety-riddled priest, Father John Patrick Mulcahy (Rene Auberjonois), and Hawkeye, framed in a medium close two-shot. The seriousness of this conversation—Father Mulcahy is tortured by Painless's decision to commit suicide and worries if he can perform the last rites knowing how his death is to occur—is what Altman deflates when his camera then zooms out to a long shot in order to reveal, first, a violinist, standing in between two nurses, playing plaintive music as if to prepare Painless for death, and then, as the lens continues its zoom-out, the larger group in the unit arranged around Painless as the disciples are, laterally, around Jesus in *The Last Supper*. But even this momentary tableau is quietly upended. Altman inflects the lateral arrangement of the figures in long shot with two cuts-in to closer shots, first to Painless and the friends immediately surrounding him, and then to two others at the far end of the table who perform a sparse variation of the film's theme song, "Suicide Is Painless." This shot complements the aesthetic effect of the tableau; the two figures performing this song are singing as much to us as they are to their fellows in the scene, as the singer stares into the lens of the camera. But the composition also fragments the tableau effect, as if in an intentional gesture to break apart any sense of painterly representation. The sequence as a whole, although culminating in a long shot enabling the viewer to read the image as a citation of a well-known painting, is built on Altman's preference to keep the widescreen frame always in a state of uncertainty and instability. This prevents his Panavision frame from ever resting with one laterally arranged view. This tendency in Altman to disperse and perpetually reconstruct arrangements within and across the Panavision frame has important consequences for any reading of figures, or figural movement, in his anamorphic cinema. We are alive to the undulating process of widescreen cinema's creation when we attend imaginatively to Altman's films; the viewer's mode of attention is a crucial part of his anamorphic project. "The movie you saw," Altman once suggested, "is the movie you are about to see; the movie you saw is the movie we're going to make" (qtd. in Smith and Jameson 166). This mode of attention is eventually self-consciously formalized and thematized by Altman, who by the time of *Nashville* becomes suspicious of the wide frame's pretensions toward enabling democratic participation in the viewer's experience.

This chapter will explore the various ways the dispersive aesthetic of Altman's widescreen cinema parallels and intersects with the worlds of his films. The idea of anamorphic Panavision as an "expansive" process in which an initially narrowed image on the celluloid strip is eventually "unsqueezed" in projection has particular metaphoric overtones in the reception of Altman. In viewing his work, we must variously move from relatively narrow points of stability to gradually unfurling and more expansive views that throw earlier assumptions into question. The first stretch of the chapter looks at relatively intimate Altman films in Panavision, many of them centered on one or two protagonists. This selection of films mainly focuses on Altman's early work, including *Brewster McCloud*, *The Long Goodbye*, *California Split*, and *3 Women*. At the end of the chapter, I look at a selection of Altman's ensemble frescoes in widescreen, exploring visual motifs in two of his large-cast, multiprotagonist movies, *Nashville* and *Short Cuts*. My reading of the latter of these two "mosaic" films is inflected by a brief sojourn to the tapestry of *The Company*, Altman's uncharacteristically emotionally warm film about ballet.

Brewster McCloud

Brewster McCloud (on which Altman collaborated with two cinematographers, Jordan Cronenweth and Lamar Boren), unlike *M.A.S.H.*, does not present its lead character, another antiauthoritarian, countercultural male—Brewster, played by Bud Cort—quite as uncritically as does the earlier film. Brewster, shy and childlike, is obsessed with birds and lives clandestinely in an abandoned security room in the gigantic and ghastly Houston Astrodome. He spends most of the film taking and studying pictures of avian specimens as he prepares to build a machine enabling him to fly across the Dome. He is also something of an unexpected, unlikely vigilante, responsible for several offscreen deaths of pungently unsympathetic figures, all of whom represent (in an unusual deployment of rather obvious social symbolism in Altman) a different social ill: avarice, racism, and misogyny. Although Brewster is the title character, Louise (Sally Kellerman), an ethereal guardian angel figure who intervenes to protect Brewster from threats outside the Dome, becomes especially crucial to the film's formal investigation of the possibilities of film editing in relation to the anamorphic frame.

Brewster's obsession with birds, part of a larger, patterned use of avian motifs in *Brewster McCloud*, involves everything from his flying machine, birds in a zoo, and bird shit left on the various corpses Brewster leaves in his wake. Louise, Brewster's apparent guardian angel, is also central to the bird metaphor, and it is she who connects it to the free-flying properties of film editing as a technique. She first visits Brewster in his Dome enclave. She asks him if he thinks much of girls or of sex. With the camera positioned behind her, Kellerman removes her trench coat—Louise's only article of clothing in the film. Her back now facing us, marks of tattooed angel wings, spread across her exposed shoulder blades and around her spine, are revealed. The shot literalizes Louise as a guardian angel and establishes her winged command over anamorphic space (Figure 3.1). For Brewster,

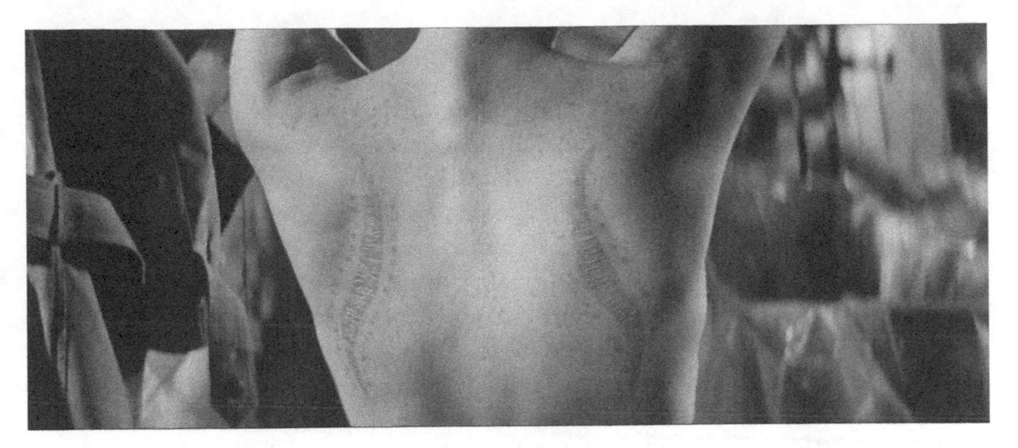

FIGURE 3.1 *Brewster McCloud* (Metro-Goldwyn Mayer, 1970). Digital frame enlargement.

Louise is not so much erotic object as maternal, protective force. Louise's goal is largely to help Brewster fly out of any potential cage thrown down around him, in both the literal and metaphoric senses of the word. But Louise's association with flight extends beyond her narrative function, generating metaphoric linkage with the very operation of film editing in the context of anamorphic cinema. In an important sense, Louise is never really part of any stable, human-centered mise-en-scène in Altman's film, or at least not subject to the same laws of movement. She can appear anywhere, at any time, her very being enmeshed with the transcendent properties of film editing, a device that shapes the form of any film without necessary regard to possible movement in the physical world. She even, at times, seems to be part of our world, or perhaps Altman's world, as in one shot in which she winks at the camera after it catches her bathing naked in a public water fountain. She will demonstrate this ability to transcend time and space most saliently in the film's later car chase, a send-up of *Bullitt* (Peter Yates, 1968) in which Louise need not pay heed to human laws of the physical world, able to appear in her automobile wherever she needs to be in order to intervene on Brewster's behalf at just the right time. But we actually see a more subtle sign of this association with film cutting in Kellerman's movements in this earlier sequence in the Dome hideaway, for her act of disrobing and pinning up her hair are two gestures that motivate two cuts in the scene: first, to a shot showing us that Brewster is uninterested in Kellerman's naked body; and, then, to a subsequent shot showing us Kellerman putting up her hair, preparing to wash Brewster in the bathtub as a mother might a child. As she bathes Brewster, she warns him to avoid romance and the traps it posits for someone focused on higher goals. Earthly love, she counsels, too often leads to conventional forms of life, to more of what humans have always already done. Brewster, she can tell, seeks more than this. With these words, and a cooing lullaby, Louise's tender warning cues a series of dreamlike shots, a travelogue in the sky that engulfs the entire Panavision frame, apparently filmed from an airplane, as if from a bird's-eye-view flying above the clouds and toward the horizon line.

Brewster McCloud ultimately inflects these expansive images, very like the point of view of a bird flying in the air, with a great deal of irony. Such images of relatively long duration—these images of dreamy flight—are unusual in this film, given its erstwhile emphasis on salient, staccato editing patterns. As Hamish Ford observes, *Brewster McCloud*, rather atypically for Altman, "features a disorienting editing style as its overarching formal principle," with "Altman's special formalism and experimental attitude towards what constitutes a feature film . . . given joyous and inquisitive expression" (125). It makes sense, then, that subject matter and formal tendencies more familiar in the intimate 1.85:1 frame would find expression, in *Brewster McCloud*, in a wider image, given the film's counternormative, experimental sensibility. Louise's association with film editing, and with the figure of the free-flying bird more generally, is also ironic in the anamorphic world of *Brewster McCloud*, with its use of a frame that would seem best suited to charting the visual movements of onscreen figures within the expansive frame, rather than transcendentally across it via disorienting cuts. Sally Kellerman's character becomes the onscreen embodiment, in this way, of the film's disruptive handling of widescreen Panavision.

McCabe & Mrs. Miller

With the relatively expansive frame granted to filmmakers via the new technology of CinemaScope in the 1950s, coolly controlling the audience's attention and response via classical editing construction became a problem. How could the viewer's eye be profitably directed when so much lateral breadth and width might occupy their attention? This potential for widescreen to engage the viewer beyond narrative is part of what informs *McCabe & Mrs. Miller*, the first of Altman's three collaborations with cinematographer Vilmos Zsigmond (with whom Altman would work again on *Images* and *The Long Goodbye*). The story is about a naive capitalist, McCabe (Warren Beatty) who arrives in the Pacific Northwest to establish a business in a town presently under construction, an effort that draws the attention of the similarly ambitious brothel madam, Mrs. Miller (Julie Christie). As in *M.A.S.H.* and *Brewster McCloud*, our conscious engagement with the materially implied process of Altman's widescreen filmmaking parallels the human feats of construction occurring onscreen—in *McCabe*, the gradual building of the town (which requires the expenditure of economic means).

McCabe & Mrs. Miller's familiar genre traits are precisely part of what enabled Altman's interest in exploring the nooks and crannies of the 2.35:1 frame in making this film. "So everybody knows the movie," Altman remarks, "those characters and the [western] plot, which means they're comfortable with it and gives them an anchor. And I can really deal with the background" (Thompson 60). But by the time Altman directs this film in the early seventies, directors working in the widescreen format had harmoniously adapted the frame shape to the conventions of continuity editing, taming the tendency of the wide image to disrupt or disorient the viewer relative to earlier, normative Academy aperture practice. What makes *McCabe & Mrs. Miller* so striking in a study of widescreen cinema is the extent to which Altman renews, in the early seventies, the anamorphic frame's potential for relinquishing authoritative mastery over the viewer's expenditure of visual attention. The film does this by using certain traits of continuity construction ironically, as a way of opening up, rather than closing down, possibilities for engaging with the wide image. This formal enabling of diversely contingent subjectivity is thematized in the film, given Altman's interest in marginal characters, who include several important supporting figures, such as Chinese workers in the town and an African American couple who live within it. Crucially, however, these implied, contingent subjectivities are never fully thematized in the narrative itself, avoiding an overdetermination of the viewer's engagement with the wide frame. The widescreen aesthetic of *McCabe & Mrs. Miller* enables an intellectual involvement with the frame loosely paralleling, without of course exactly replicating, the efforts of these late-nineteenth-century characters to forge their own path in a new American town.

As Robert T. Self has commented, Altman, along with one of his central collaborators on this project, cinematographer Vilmos Zsigmond, works throughout the film to "sublimate logical connections to the stasis and imagery of poetry" (*Robert Altman's* McCabe & Mrs. Miller 136). This sensibility is evident in Altman's recurring, busy plenitude of onscreen and frequently contingent movement in his wide frames. But *McCabe & Mrs. Miller*'s overall construction as an anamorphic work also delivers the impression that it is a film always perpetually starting over, restarting itself, even as the narrative ostensibly proceeds in each moment with a new image, a new arrangement, a new tableau. Robert Niemi has noted this quality, observing that *McCabe* is "structured concatenatively, by a series of arrivals" (40), a structure appropriate for a film about characters engaged in architectural construction and entrepreneurial ambition. Niemi characterizes these recurring arrivals in terms of the gradual entrances, one after the other over the course of the film, of the major figures: McCabe, then Mrs. Miller, then the corporate business representatives, and then the villains hired to gun down McCabe after he turns down the representatives' offer. But in addition to these arrivals in the narrative, Altman uses a repeated visual conceit, structuring many sequences around the arrival of one of the major or supporting characters (or even an extra) to a particular space—the saloon, the brothel, the church, or some other part of the growing town or the landscape surrounding it. These arrivals generate a sense that we are perpetually, again and again, being brought into slightly different spaces, or fresh views of spaces we only thought the wide frame had fully displayed earlier.

Altman orchestrates his anamorphic frames in this way for several reasons. First, it is rarely the arrival itself that is the salient point of interest in a sequence.

The arrival of an onscreen figure (walking across a landscape, or through a door, or across a bridge) is a way of conventionally drawing us to a new slice of screen space but without then analytically breaking down that space through conventional means. The arrival usually justifies, instead, a perceptually fresh placement of the camera by Altman and Zsigmond that makes a space we have already seen strange and new not only because the town continues to expand as the film goes on but also because the anamorphic frame has been composed in a different way for every renewed view of an ostensibly familiar space. *McCabe & Mrs. Miller* is a film full of laterally arranged and composed establishing shots that, in fact, do not very often establish new spatial relations meant to be broken down through a procession of sequentially closer shots, as in conventional analytical editing. Instead, these poetic shots in *McCabe* signal transitory, momentarily held ways of occupying screen space shortly turned over to another, different moment of held stasis in which our attention is fixed by some other fascination of light, movement, or gesture. This aesthetic is thematically suitable for a film that is itself about the gradual transition of a small town currently undergoing—in the Old West, a highly contingent and extremely dangerous—process of being built; the town, for the duration of the entire film, is gradually establishing itself, and so it makes a certain amount of narrative sense for the establishing shot imagery to function in this unexpected way. Experientially, however, this aesthetic works less to allow the viewer to consume straightforward narrative information, that is, to convey the relatively sparse and redundant information that "the town is now establishing itself, and here is McCabe again, working to establish his business." Instead, such moments of relative stasis engage the viewer's eye in creative and relatively underdetermined ways, a poetic correlative for the (eventually brutally circumscribed) efforts of some of the characters to eke out a way of living.

The film's tendency to use the widescreen image to repeatedly "establish" a space we thought we had already become familiar with is perhaps no better illustrated than in the film's violent, but also painterly and languid, denouement. McCabe has refused the offer by the businessmen to buy his properties in the town, and he is now hunted by a trio of villains hired by the company, led by the burly Dog Butler (Hugh Millais). Seeking a view of where the gunmen might precisely, presently be, McCabe retreats to the tower of a church, pictured by Altman in one of the film's memorable extreme long shots, with Beatty internally framed in the window under the church steeple against a bright white but also forbiddingly sparse sky. A cut to what is ostensibly his point of view from this vantage point shows, for the first time in the film, the expanse of the town, now seemingly fully built—with the exception of this very church, which, significantly in terms of the film's melancholy vision of the American West, is still hollow on the inside. If there is anything like a magisterial establishing shot in *McCabe & Mrs. Miller*, this is, at least on the surface, it: we see nearly the entire expanse of the town laterally arranged across the widescreen image for our momentary perusal. But despite the apparent mastery of the image, this is also a very subjective vision: McCabe's momentary, and in the face of his impending death, highly contingent point of view. McCabe is not in a position of analytical mastery over this terrain, a position he has arguably sought throughout the film through his desire to attain more and more capital; the gunmen coming for him remain threateningly out of view. Also creating a problem for McCabe—both in terms of his ability to see the three gunmen and in terms of his ability to physically move through this landscape—is the fact that snow is starting to fall across the town, Altman's images becoming gradually blanketed in piles upon piles of white snow obscuring the precise spatial layout of this new town's now-complete construction. The overall effect aligns our own point of view with the discretely situated, rather than masterfully transcendent, McCabe, not in a psychological sense but in a purely pictorial one. A town, built through the efforts of small, and not always entirely professional, businessmen like McCabe, becomes blanketed by snow, dotted by blood.

As if to underscore the way in which his wide frames are always already undergirded by a contingently situated, dispersive, nonmasterful subjectivity, the entire expanse of the frame, after McCabe dies, eventually becomes engulfed by wafting clouds of smoke, and by Julie Christie's eye, as she drifts into sleep in an opium den in the film's final shots. Her indulgence has the effect of now sending the film's erstwhile depiction of physical terrain through the filter of an eye—an eye that now engulfs the entirety of the anamorphic frame—as if through a catatonic sieve.

Gains and Losses in Altman's Films on Anamorphic DVDs

A peculiar self-awareness of the widescreen frame is always potentially available in viewing letterboxed DVDs, Blu-rays, and other physical home media preserving the original theatrical aspect ratios of films. This awareness of the anamorphic frame is enabled by the presence of black bars when 2.35:1 films are displayed on modern 16:9 HDTVs, bars thrown into relief against the ratio of the television set itself. In such a diminished viewing situation, the wide frame unwittingly becomes vulnerable to an unintended irony, as if the letterboxing bars were quotation marks now appearing over formerly, earnestly expansive images. Although home theater discourse always threatens to devolve into class-based, conspicuous consumption of movies as luxury products—meant to be shown on, and to show off, the home theater owner's audiovisual apparatus and concomitant luxury status—extrafilmic commentary and extras on DVDs and Blu-rays still carry the promise to restore to widescreen discourse some of the more substantive pedagogical, intellectual, and aesthetic dimensions of anamorphic cinema.

Murray Pomerance, in his essay on *The Long Goodbye*, points to the pedagogical value of one such extra on the MGM DVD release of Altman's film. In an interview on a documentary extra on that disc, Altman discusses the film's self-reflexive use of the John Williams and Johnny Mercer jazz theme, and the way in which five notes of this music are used, diegetically, as the sound of the doorbell of the house belonging to Roger Wade (Sterling Hayden) and his wife, Eileen (Nina van Pallandt), heard after private detective Philip Marlowe (Elliott Gould) presses it. Pomerance draws our attention to Altman's pleasure, in the included interview on the DVD, in pointing out this diegetic play with the film's nondiegetic theme song—a stylistic gesture that a less attentive viewer, without benefit of Altman's commentary, might have missed. For Pomerance, this interview suggests that Altman "stands outside the film and the viewing audience, but above, occupying a position from a high vantage, taking an all-inclusive view" ("High Hollywood" 235). And in providing this citation of his own filmmaking trick for the viewer in this documentary extra, Altman is also inviting the attentive spectator to make use of this knowledge in their own play with the possible meanings of his film. Here the substance contained in a documentary interview with Altman on the DVD constitutes a kind of discursive framing of the film's own frames, inviting the viewer to entertain their own thoughtful view of Altman's widescreen compositions, preserved via the letterboxed DVD. Such a bonus feature carries pedagogical value rather than the emphasis on consumerist value so prevalent in discussions of home theater technology.[1]

And yet even as the discursive surround created by physical media generates intellectually substantive pedagogy, viewing widescreen cinema on DVDs can also introduce perplexing obstacles and palpable losses. The second major post--*M.A.S.H.* collaboration between Altman and Elliott Gould, *California Split*, was transferred to DVD in 2004, its first release on any home video format. The cut contained on that disc, however, was significantly truncated due to the inability of the producers to secure home video copyright for all the songs featured in the original theatrical cut of the film. The commentary track to the DVD (which includes the participation of Altman, the film's screenwriter, Joseph Walsh, and stars Gould and George Segal), while replete with intriguing observations, also contains one moment of apparently unintentional irony. Walsh, for a stretch of the commentary track, talks about the film's clever play in blurring the lines between diegetic and nondiegetic music in a late scene in which Gould, walking across oncoming traffic in the streets of Reno, sings lyrics from the song "Me and My Shadow." The viewer *sees* this scene on this DVD, properly preserved in

[1] The best recent historical example of this type of discourse about home theater technology, in print form, is the California-based magazine *Widescreen Review*, which from its November/December 1992 issue to the present day has served as a reliable consumer's guide to home theater technology. Its numerous reviews of movies transferred in widescreen form to home video (in the early years of the magazine, on laserdisc, and in recent years on DVD and Blu-ray) focus not on questions of aesthetics or film criticism but on an evaluation of the transfer of the film image on various video formats as well as the audio quality of the video transfer. The early issues of the magazine featured fascinating and lengthy articles on aspect ratios in cinema and the relationship between celluloid-based widescreen imagery and its various home video transpositions, discourse worthy of further study.

the original 2.35:1, as Walsh talks about it, but does not in fact hear the music to which Walsh refers. Walsh, talking over the scene and likely working from his memory of the film, probably did not notice that the song was in fact missing from the cut on the disc for which he was providing commentary, its having been cut from the DVD version with only an instrumental tune accompanying Gould's stroll (see Gayle Magee's essay on *California Split* for more on the implications of this and other missing tunes from the film). Such a discordant moment of commentary signals the ironic loss of the original, widescreen theatrical experience even as DVD and other home media technologies are ostensibly positioned to preserve it. But even this loss, like Altman's clever citation of his filmic trickery in *The Long Goodbye*'s DVD supplement, can cue our own playful engagement with Altman's frames, especially given that loss is a major theme in *California Split* itself.

The Long Goodbye

In *The Long Goodbye* (Altman's second collaboration with cinematographer Vilmos Zsigmond), Philip Marlowe's signature wardrobe of white shirt and black jacket and tie stands in vivid contrast to the overexposed pastels of the Los Angeles landscape and suggests a deductive consciousness distinguished from the cacophonous city surrounding him. *The Long Goodbye* floats across Los Angeles jetsam, surveying a plenitude of visual information that does not always result in narrative (for us) or deductive (for Marlowe) substance. The impersonality of Los Angeles is registered by the film not only through the horizontal compositions that frequently keep the city in the background but also through Marlowe himself, a self-contained, singular, apparently honest man who contrasts to the morally questionable world surrounding him.

In addition to being a kind of progenitor of later neo-noir, *The Long Goodbye* is also a knotty private detective film, employing its mostly restricted narration to keep us by Marlowe's side throughout nearly all of the movie as he investigates the apparent death of Sylvia Lennox, the wealthy wife of Marlowe's close friend Terry Lennox (Jim Bouton). Altman's take on the technique of restricted narration is informed by his use of the anamorphic frame. There is, to take a striking example, at least one moment in which Marlowe is not present to the precise events we witness: a conversation between alcoholic writer and possible murderer Roger Wade and his wife, Eileen. During this moment, Marlowe is instructed by Wade to leave them alone for a minute, to go walk along the beach while they settle a domestic spat. But Altman keeps Marlowe, out of earshot along the shore, nevertheless within shot of the viewer's eye, in the reflection of the glass door pane, a strategy in keeping with several presentations of Marlowe in relation to Wade and his wife throughout the film, in which Marlowe is presented at a distance from these other figures (Figure 3.2). These shots suggest that Altman prizes visual engagement with his primary character, and with the world of a character, above and beyond a strictly literary conception of restricted narration,

FIGURE 3.2 *The Long Goodbye* (United Artists, 1973). Digital frame enlargement.

just as Marlowe knows that keeping oneself at a distance from the players in the case is at times a necessary if inconvenient move for the private detective as he journeys along the complicated trajectory of divining whodunit. There is, at nearly any moment throughout *The Long Goodbye*, all manner of clang and clamor in and around the anamorphic frame competing for our, and Marlowe's, visual and aural attention—a hungry tabby cat, mewing for a particular brand of cat food; the distant chatter of a group of young women, variously naked, across the way from Marlowe's residence; and the honks and tire squeals of oncoming traffic as Marlowe attempts to catch up with Eileen's car near the end of the film. Such a plenitude of Los Angeles clatter makes ratiocination in *The Long Goodbye* especially difficult and our own act of filmic perception open to divergent, and not always rationally relevant, details.

Isolated from the noise of the city near which Marlowe lives, Roger Wade's luxurious beach house (Altman's own Malibu home at the time the film was made) is a location in which Altman composes complexly mobile anamorphic compositions. When Marlowe first arrives at the Wade home, he smokes a cigarette and ambles around the room, waiting for Eileen to finish a phone call. The camera, loosely approximating Marlowe's point of view, begins a gradual zoom-in toward a glass case displaying the preserved body of a dead bird, the reflection on the glass a view of Eileen, from across the room, talking on the phone. Deception is already in the air given that Eileen has falsely introduced herself to her telephone interlocutor as Wade's secretary, rather than his wife, and with the dead bird suggesting an unpleasant fate for anyone, such as a private detective, who has the pretension of taking on an authoritative view of things. As a perception given to the audience, this image of the reflection and the bird is nevertheless ambiguous; the first time I saw the shot, I focused on Eileen's reflection rather than the bird, which did not become legible to my eye until a subsequent viewing. The frame in this way opens up the possibility that the viewer will engage with the anamorphic frame—even in a shot such as this one, ostensibly an image corresponding to Marlowe's approximate point of view—differently from the protagonist, a gesture toward the diverse ways of seeing within and inhabiting the widescreen world that will become even more intensified in Altman's later, large-cast mosaic films. Marlowe has noticed the bird—he makes a quip about it to Eileen after she joins him at the end of her phone call. If we get the joke, it means our fixation on the reflection of Eileen has not been so absolute as to preclude our noticing information that does not seem to have much of anything to do with solving the case. Marlowe opens himself up to the extraneous details within and across Altman's frame not because he necessarily thinks it will fit into a solution to the case of Sylvia Lennox but simply because this is the surreal, multitudinous world in which his skills of focused ratiocination must wittily thrive.

Altman rarely uses his widescreen frames to create overwhelming or sublime moments in which the frame is momentarily stifled by the engulfing, indeed immersive, experience of *one* primary sound, *one* primary image. Yet such a sublime stretch of cinema occurs, most unexpectedly, in *The Long Goodbye*, in the sequence in which Roger Wade, who we will soon learn—but not exactly in this sequence—is guilty of the murder of Sylvia Lennox, commits suicide in the Pacific Ocean. At the beginning of the sequence, Roger, utterly stupefied by alcoholic intoxication, is put to bed by Marlowe and Eileen. Eileen invites Marlowe to stay for dinner. Eventually the sequence brings us to a zoom-past shot of Marlowe and Eileen, standing next to a glass door, conversing about Wade's potential involvement in the case. As Marlowe and Eileen converse, the zoom lens takes us visually past them and toward waves crashing against the shore, where we eventually glimpse Wade drunkenly heading toward the water. As the zoom-past draws our interest toward him, Marlowe asks Eileen a key narrative question: Where was Roger Wade on the night of the murder? But we hear not her answer but the overwhelming sound of the ocean waves as the camera cuts to an exterior view of Marlowe and Eileen speaking by the window, the sound of the waves now drowning out their conversation. A pair of shots of Roger Wade drowning himself now commands the attention of the entire expanse of the widescreen frame, and in turn commands the attention of Marlowe and Eileen, who in a shot sandwiched between these images of Wade drowning himself notice, from their perch by the window, that he is presently committing suicide. The engulfing images and sounds of the ocean that swallow up both Roger Wade, and perhaps the solution to Marlowe's case, are interrupted only intermittently, and then dimly, by the sound of the Wades' barking dog and the pleas of Marlowe to Eileen to give up her search for Wade in the water. In these shots of the engulfing water there is nowhere else to look but into the eternal ocean into which Wade has flung himself. Even here, however, Altman focuses on the liminal, hidden details in the frame, which in this case it takes a dog to discover: Wade's cane, left behind on the shore after his final stroll into the water, retrieved by the Wade family pet less affected by the sublime sounds of the fatal ocean that swallows up the film and which occupy for a striking moment our, and Marlowe's and Eileen's, visual and sonic attention.

California Split

California Split, the first of Altman's three films with cinematographer Paul Lohmann (to be followed by *Nashville* and *Buffalo Bill and the Indians*), is about a friendship between two gamblers: devil-may-care, live-for-the-moment Charlie (Elliott Gould) and buttoned-up, debt-ridden William (George Segal). The film generates tension between the joie de vivre of the compulsive gambler, living on the knife's edge of chance, and the melancholy fate awaiting after the thrill of the moment is gone. As in other Altman films such as *Nashville* and *A Prairie Home Companion*, music becomes an important resource to the film and its exploration of the gambler's life and losses. Gayle Magee closely analyzes the original theatrical cut of the film and its use of various jazz, pop, and Tin Pan Alley songs, sung not only by Gould and Segal but also, in a way that blurs the distinction between diegetic and nondiegetic music, by Phyllis Shotwell, a lounge singer and pianist who performs at the Reno casino in which William finally wins big. As Magee notes, we hear Shotwell's singing throughout the film, and during a first screening we take these songs to be nondiegetic performances, scoring various stages of Gould and Segal's friendship, before it is revealed that Shotwell herself presides over the piano in the Reno joint inhabited by the two gamblers in the film's final stretch (223). Her songs are a key part in the film's structure. They function as a kind of siren's call leading William and Charlie to Reno, suggesting that the freewheeling behavior and lifestyle of the two gamblers is governed by a larger structure or force.

In a way that slightly reverses the apparent discursive gains about widescreen in the home theater environment via Altman's insightful interview on *The Long Goodbye* disc, however, not all of Shotwell's songs are audible on the DVD edition of *California Split*. As mentioned earlier, several had to be removed for the film's home video release due to problems with copyright. In the case of this particular DVD, the restored widescreen presentation on home video is enabled by loss rather than full preservation of the original theatrical film, losses that include the removal of music scoring some scenes and, in one case, the truncating of minutes of screen time in which particular cut songs were originally heard in the theatrical cut. The same copyright laws, at the time of this book's writing, do not apply to streaming versions of films, since streaming licenses are considered by the studios to be like television broadcasts and not equivalent to physical media. For this reason, the original theatrical cut of *California Split*, complete with all songs, has occasionally been available on streaming services such as Amazon Prime Video and the Criterion Channel during my time of writing, affording the opportunity not only to hear the cut music but also to see the imagery from the original theatrical release excluded from the DVD because of soundtrack issues. Ironically, it is in this case not physical media generating discourse about widescreen cinema but rather the elusive, temporary world of a film's streaming availability that affords a fresh opportunity to learn about what these cut scenes from *California Split* can teach us about Altman's anamorphic practice.

The most substantial footage cut from the film appears in the final Reno sequence, in which Shotwell sings "Georgia on My Mind." On intact versions of the original theatrical cut, the song and the approximately two minutes of footage it scores are restored. These images constitute a series of shots of Gould and Segal, sitting in separate spaces in the casino, cut in parallel. William plays blackjack while Charlie—following William's instructions to leave him alone so he can concentrate on his game—kibbitzes at a bar near Shotwell's piano in the casino. Shotwell's spirited, smoky rendition of "Georgia on My Mind" wafts in on the soundtrack in accompaniment of a series of tightly composed images: close shots at the blackjack table, of Segal's hands as he doles out his bet and considers his cards; a tilt-down from a close shot of the dealer to the tag on her blouse, which reads "Barbara," continuing a motif in this film whereby a comical number of female characters are so named; a medium close shot of Shotwell at her piano, playing her song, and then a zoom out to reveal Charlie still kibbitzing with other gamblers at the bar; a cut back to several shots of William at the blackjack table, smiling as he wins; and a shot of Charlie tipping Shotwell with some cash as she finishes her rendition of the song. If, as Gaylee Magee observes, Shotwell's songs have throughout *California Split* worked in part to create "an internalized lounge that these addicts carry with them whether they are actually gambling or not" (221), in this sequence that internalized space finds its externalization in a red-velvety casino joint in Reno. However, the restored shots in the streaming theatrical cut during Shotwell's rendition of "Georgia on My Mind" underscore the loneliness of this abode, an activation of the wide image that for the characters results in no wider expanse of sociality. William, having sent Charlie away to focus on blackjack, only feels most at home as a gambler when he is alone and able to concentrate on his playing. After winning big at cards, William will reunite with Charlie (in a sequence present also on the DVD version) at the craps table, where the two of them continue to rake in winning chips. Altman keeps us relatively close to both Charlie and William during these shots, focusing on their apparent camaraderie and intense concentration as they place their bets and roll the dice. But these intimate moments result in no corresponding celebration in the wider frame, even after the pair win big. When Altman does cut to longer shots that populate the wide frame with lookers-on in the casino, these images only underscore the extent to which the curious attention of others toward William and Charlie is fleeting and insubstantial.

The film suitably ends on a note of aloneness, now minus the rush of elation briefly accompanying Charlie and William as they have won these games, reinforcing the solitude that pervades the film. As Charlie goes to the cashier to cash in, William sits alone at a table, exhausted and despondent. This image of

FIGURE 3.3 *California Split* (Columbia Pictures, 1974). Digital frame enlargement.

William, internally framed and alone in a separate room while onlookers watch Charlie cash out the winnings, is one of several such frames of William alone in the movie. However, his solitude is now further inflected, and ironized, by the offscreen clamor, the noise of poker chips and slot machines (Figure 3.3). "There was no special feeling—I just said there was," William confesses when Charlie brings the piles of cash into the room. And yet despite the film's emphasis on William's feeling of emptiness, these frames are not determined by any kind of anti-gambling moralism; even when his frames are absent melodrama (as they usually are), Altman does not fill up the wide space with a strident message. "And yet that same emptiness," after all, as Wheeler Winston Dixon has written, "also *propels* the film, as Bill and Charlie search for something, anything, to give them reason to keep on living, although such a darkly metaphysical question would hardly occur to either man" (176). The irony of the use of the anamorphic frame in *California Split* is that the apparent expanse it offers, the lateral space in which to imagine other ways of living, can only be filled by these men with the roll of the dice or a bet on a game.

3 Women

3 Women opens with a complex sequence that suggests connections between the act of creating a painting—a gesture of artistic craft that will repeat throughout Altman's later work—and the act of looking at a pictorial frame with no apparent or obvious finish point, no clear border demarcating it from the larger world. The first shot of *3 Women* (the first of two Altman collaborations with cinematographer Charles Rosher Jr., who would also shoot *A Wedding* the following year) pans toward the left, from behind a diffuse and impressionistic layer of fog, toward the out-of-focus image of a painter, apparently female, facing away from the camera, creating a fresco. Completing its movement left, the camera, now zooming out, begins a return rightward, the frame expanding to reveal more of the abstract, labyrinthine painting being created in front of our eyes. Altman's own composition emphasizes desaturated beiges and grays alongside the cool blues and subdued reds of the painter's fresco, but the precise content of this frame, which appears to be generating itself in rhyme to and in tandem with the gesture of the painter, is less important at this early moment in the film than our perception of it. What appears to be water first trickles and then floods nearly two-thirds of the frame in a deeper blue before disappearing again, giving way to the title card: *Robert Altman's 3 Women*, rendering Altman's authorial inscription across the wide frame. The blue water near the bottom line of the frame appears again as the camera continues its rightward pan to reveal more of this painting—which consists mainly of reptilian figures variously equipped with large breasts and gigantic penises. What is also becoming slowly apparent through this camera movement is that this painting, although of the representational variety, is hardly stable: although the anamorphic frame would seem to be the ideal format through which to view this expansive, perpetually expanding fresco—itself a work in progress in the diegesis—Altman disperses any sense of stable perception, perpetually refiguring these painted figures through a mobile frame and through the occasional intrusion of blue water from below the bottom frame line, repeatedly flooding the image with aqua.

Altman will end this shot with a slow zoom-in to the painting itself. This gesture of moving the camera closer to a representational work, to get a better look at its textures and lines, eventually gives way to another abstraction: a dissolve to another shot, the out-of-focus blue of what will eventually be revealed to be a swimming pool. This play with abstraction that will come to define the use of the anamorphic frame in *3 Women* is here rudely interrupted by the sudden presence of an obese, aged pair of legs, walking into the pool, a ruinous figure at odds with the eternal permanence, and robust and assuredly sexual physicality, of the painted figures on the fresco. These are the legs that move us, wobbly, further into narrative: this space will eventually be revealed as the geriatric center where the film's two main characters, physical therapists Millie (Shelley Duvall) and Pinky (Sissy Spacek), work. But Altman keeps this narrative information at bay for a stretch, continuing to occupy this series of shots with abstractions of water and bodies, and with the same slow, viscous engulfment of the Panavision frame. After Altman's writer-producer-director credit, the final shot of this title sequence ends with one of the more jarring cuts in this filmmaker's anamorphic cinema. From this pool of fragmented bodies and water drifting centrifugally and perpetually out of frame, the film transitions to a shot, from a longer distance, of recognizably human figures going about their business, a group that includes the woman who will be one the film's two main protagonists, Duvall's Millie. This collection of bodies arranged across the frame would not be out of place in any one of Altman's later, multiprotagonist frescoes, with its emphasis on diverse human movements and perspectives in the frame. The film has just begun, and we do not yet know Millie as a character, although fans of Altman will be able to pick out Duvall in this image. But the camera's movement rightward, past the pool and to the other side of the room, eventually leaves Duvall behind.

Just as jarring, but in a quieter way, is this moving frame's eventual revelation of Sissy Spacek, as Pinky, standing behind a window, looking out onto the pool. Where Altman's mobile camera and dispersive arrangement of the Panavision frame has, up to this point in *3 Women*, emphasized a panoply of impressions and points of abstract interest, suddenly now an ostensibly stable center is established, a single point of view—although in the case of Pinky, it will turn out, an especially obsessive and, as soon becomes clear, quite unstable perspective. Spacek's intense act of looking here is something of a bookend, in this first stretch of the film, to the gesture of the painter with which the film just a moment ago began. If the gesture of the painter inaugurated the film's initial play with figuration and its ensuing dispersal through the movement of the wide frame, here Spacek's gaze seems to imply that looking in Altman's film has its own aesthetic implications and effects. Typically, in cinema, the gesture of the actor and the gaze of the viewer are situated as different entities. But Altman conflates gesture (the painter's creation of the tableau in the opening shot) and looking (Spacek's intense gaze), granting both gesture and gaze aesthetic energy within and across the anamorphic frame. The film expands the possibilities of what might be considered aesthetic labor, locating in the very act of watching—our watching of these frames, rhymed with Spacek's gaze—a potential source of creativity, an optical gesture intimately linked to the painter's brush.

In contrast to *Brewster McCloud*, in which the exuberant and eccentric formal operations of the film itself are expressively conjoined with the eccentric powers of Louise, the birdlike guardian angel, in *3 Women* the film's own creative, aesthetic play, initially associated with Pinky's way of seeing, gradually becomes severed from the subjectivities of the characters. The anamorphic frame in this film instead develops a motif whereby it will move past the characters, slipping beyond them to an unwelcoming, arid world. Robert Kolker has commented on this effect of the film repeatedly losing from its view its already lost characters, a narrative strategy denying the viewer any clear point of identification. "Pinky is filled by Millie," Kolker writes, "for she has no self at all." He goes on:

> Later, at the spa, when she talks incessantly about Millie and how she misses her, the camera slowly and deliberately zooms to and past her face—achieving an effect akin to having her face slide slowly off the side of the Panavision screen—to the bright windows behind her, which go out of focus. There is an immediate cut to a metered television set on the wall of Pinky's dark, closed room. The set is on, but there is nothing on it, and the camera pulls back from the bright empty screen (which suddenly goes off) to Pinky lying asleep. (391–392)

Kolker effectively describes how *3 Women* becomes an ironic manifestation of expressionism in widescreen cinema, its imaginative visuals expressing apparently blank subjectivities, empty of content. This perceived emptiness is intensified not only by the use of the wide frame, its horizontal expanse a constant counterpoint to the two main characters who do not have anything particularly meaningful to do or anywhere especially important to go within or across it, but also by the aridity of the desert landscape surrounding the California locations. Altman has taken the strategy of refusing the viewer an easy path to psychological or emotional immersion in a widescreen world to a fascinating extreme in *3 Women*.

The film is something of an allegory for a foreclosed desire for psychological and embodied immersion in a world in which the very possibility has become nothing but a consumerist, narcissistic cliché. Midway through the film, after Pinky has been scolded by Millie for interrupting her tryst with the owner of the apartment complex, Pinky, in a desperate and melodramatic act, throws herself into the apartment complex's pool. Although an apparent attempt at suicide on the level of the narrative, on the level of the film's aesthetic operations, Pinky, as a screen figure, is seeking to restore her connection to the film's subliminal, painterly aesthetic, a connection she seemed to possess in the opening frames, where her gaze was so intensely connected to Willie's gesture of painting. The walls of the pool into which Pinky throws herself are decorated with the same reptilian figures Willie paints earlier in the film. Right before Pinky's jump, Altman engulfs his frame with the pool's water, with certain of Willie's painted figures visible through the pool's gently undulating surface. Pinky's jump into the pool and Altman's filling of the anamorphic frame with water are pictorial representations of the desire for immersion, for engulfment, a pictorial depiction that nevertheless does not deliver to the viewer immersive effect. Likewise, Pinky's subjectivity is expressed, but its emotional resonance is severed and short-circuited by Altman's recourse to aesthetic distance, a gesture we see again later, near the end of the movie, in *3 Women*'s highly abstract dream sequence, a vivid and painterly portrayal of character psychology across a series of expressionistic frames, but one that avoids immersing the viewer into a legible emotional experience of those subjectivities.

The initial aesthetic playfulness of *3 Women* eventually gives way to its depiction of a world in which play does not offer satisfaction. In fact, it appears to be a world in which the absence of play, or the player's inability to find another to join in her games, can lead only to suicide, or to rigid positioning in predetermined allegory. The arid ending of *3 Women* finally figures Pinky, Millie, and Willie as a kind of symbolic "family without men," but absent any sense of generative or productive play. In Altman's multiprotagonist fresco films, discussed in the final stretch of this chapter, we will meet characters who attempt to restore some sense of play, and by extension a connection to Altman's own formal operations, through their art and through their craft.

Widescreen Frescoes

The wider canvas of Altman's widescreen mosaics would seem to offer a stage on which his characters, some of them engaged in their own creative and artistic pursuits, might more meaningfully intersect with the aesthetic experiences of the viewers of the films. This is an especially vivid possibility in a trio of Altman's multiprotagonist, widescreen frescoes. In these films, characters make various forms of art: country pop songs (*Nashville*), dance (*The Company*), or painting (*Short Cuts*). But more often than not, the social milieu swirling about them in the wide frame posits its own share of attendant difficulties, including the uncertain, or at times noisy and indifferent, popular reception of dedicated artistic labor. The ambiguity surrounding the reception of the work of Altman's artist characters parallels our own viewing of the films, which do not instruct or guide our reception of the art-making going on in the narratives, even as the wider frames give us a plethora of views through which to potentially make sense of the characters' creations.

Nashville

On the commentary track to the DVD of *Nashville* (his second collaboration with cinematographer Paul Lohmann), Altman observes that native Nashvillians who saw the film in 1975 reacted negatively to its music, claiming Altman's vision was an inauthentic depiction of country music culture. The film, a story about a political campaign's intersection with celebrity performance, is a parade of "country" songs written almost entirely by the film's cast, a sonic interpretation of the American South by Hollywood. Altman embraces this inauthenticity and channels it into the film. On the commentary track, Altman states that he believes some of the songs in the film are quite possibly good, others quite possibly very bad. He does not state his evaluation of any particular song, however, leaving it to the viewer to decide, much as the characters must do in the film, what to think of this music. But this "open" presentation of the songs in *Nashville*, free of judgments of taste imposed by the filmmaker, is nevertheless marked by fatalism. If one's judgment is that, say, the songs of Keith Carradine in the film are superior to those sung by Karen Black (or vice versa), it is still true that at the end of the film the country singer Barbara Jean (Ronee Blakley) will be shot by a drifter assassin, Kenny (David Hayward), regardless of whatever distinctive paths our tastes in viewing and listening have forged in an experience of the film. The ending of *Nashville*, a grim portrait of the overdetermination of American politics, is always the same regardless of taste preferences in music guiding us through a contingent experience of the film. In this way, the expansive frame in *Nashville*, ostensibly welcoming any opinion about the songs sung by its characters, carries an ironic, bitter inflection. When it comes to politics, *Nashville* implies, the particularities and distinction of our taste in the arts, in the context of an anamorphic frame that always seems to contain a cacophony of views, simply do not matter.

Throughout the film's music performances, Altman's wide frame insists on the (usually ambiguous or at least multiple and diverse) reception of the music by the diegetic audiences in the film who attend the various performances by the celebrity characters. As Charles Warren points out, "Audiences are volatile in this film. . . . Film itself is an audience, to the singing performers and to all the others in the film, appreciating them, bearing down on them, holding dangers for them. Clearly audiences are a worry to Altman—the audiences to art, and art itself in its capacity as audience" (30). These performances for which the film's camera serves as an audience are presented matter-of-factly, unstylized and without overt authorial commentary. Whatever Altman's own taste for music is, it does not shape these frames of *Nashville*. The film's shot compositions repeatedly return to the performer after making a visual gesture toward the reception of the music. The space between the performer and the reception of the performance (occasionally inscribed in the same shot, at other times connected via a combination of zooms or cuts) is not one of unity or connection (reminding us again of Warren's observation regarding the volatility of some of the audiences in this film). Nor is Altman only creating here through his wide frame a "democratic" space fillable with our own varied responses to the music (we *can* fill it with musical taste judgments, of course, but that will not lead us anywhere). Altman is intent to show in the composition of these frames the vacuum between the performance of art and its fragmented, inchoate reception in American culture, even as the political banners and waving American flags that occasionally engulf the entirety of Altman's wide frame falsely imply some unified perspective through which to make sense of clamor.

The performance of Haven Hamilton (Henry Gibson) of the song "200 Years" (written by Gibson) across several shots in the film's opening titles sequence sets this motif of uncertain reception into motion. The song is a rather pallid paean to American industriousness. For Hamilton the strength of America as a country is indicated less by any achievements in the culture and more by its sheer lasting power ("we must be doing somethin' right to last two hundred years," he repeatedly intones in this song). The song, in other words, appears to offer satire (unwittingly so by Hamilton, intentionally so by Gibson), yet the anamorphic shot compositions render ambiguous the song's reception. The first shot in Hamilton's recording studio begins not on the singer himself—although he is already audible offscreen on the soundtrack—but rather with a slow pan rightward across the studio, beginning first on a drummer before panning past a room of the guitarists and keyboardists accompanying Haven. Behind them can be glimpsed a room of listeners attending the session, seated behind a pane of glass. As the pan rightward continues, however, these musicians and listeners are in turn relegated to offscreen space. The pan incrementally brings us to more musicians—a pianist played by Richard Baskin, two more guitar players, and a quartet of backup singers isolated in a separate singing booth. These backup singers briefly share the screen space with Haven, now revealed in a separate booth on the

FIGURE 3.4 *Nashville* (Paramount Pictures, 1975). Digital frame enlargement.

right side of the frame. But they, too, are soon relegated to offscreen space as the camera's pan settles in for a close shot of Gibson, now the only figure onscreen and internally framed by the black vertical lines of the recording booth (see Figure 3.4, three moments in this laterally mobile framing).

This lateral pan across the recording studio is complex and suggestive of the film's attitude toward the production and reception of music in American culture. The mobile wide frame gestures toward the collaborative nature of the production of music, guiding us across a room full of singers and musicians working together toward the recording of a song. That this image should also contain listeners—isolated from the recording booth, but visible, behind the pane of glass in another room—suggests the extent to which this song is reflective of their desires, the listeners toward whom Haven targets his music. But even as this shot surveys a complex circuit of collaborative production and varied reception, the gradual pan, in its eventual eschewal of the audience and other singers and musicians to focus for a stretch of several seconds on a close-up of Haven Hamilton, also suggests how artistic and ideological complexity is often left behind in American culture in favor of a focus on individual celebrity figures. Once again, Altman's complex handling of the widescreen mobile frame and its lateral staging arrangements is ironic. His lateral, mobile framing depicting this recording studio would seem to emphasize artistic collaboration. But as the sequence ends, through the direction of the frame's movement, we are focused rather on a celebrity who is a figure of fetishistic fixation for some in the film's world. What is being created here in this recording studio is not so much a work of performance intended for a wide audience as a carefully calculated celebrity product delivered to one isolated demographic. When the outsider Opal (Geraldine Chaplin), a faux-BBC reporter, intrudes into the recording studio, Haven interrupts the song to order her, a figure foreign to the intended audience for this music, out of the room. Later, he will issue a similar order to "Frog" (Richard Baskin), the long-haired piano player whose performance dissatisfies Haven. Despite the expansive frame through which Altman depicts it and despite Haven's paean to America ("we're all a part of history!"), it is clear that the consumer product being created in this recording studio is calculated to appeal to only a very narrowly defined group, even as *Nashville*'s own wide frame, in its very unwillingness to inscribe a predetermined moral judgment—or judgment of taste—invites a greater range of response.

Altman's shot compositions and rhyming edits during later musical performances in the film deepen these implications. About an hour into the film, *Nashville* depicts a handful of performances at the Grand Ole Opry, staged by Altman for the film and featuring actors from the movie, but with many native Nashvillians in the background audience, their presence intensifying the faux-documentary effect of the sequences. The first—and, for this listener at least, most memorable—of these performances is by Tommy Brown (Timothy Brown), an African American country singer. Altman's strategies in this sequence are different than in the sequence of Haven recording his song in the studio. Where the earlier sequence's long take of the studio musicians complementing Haven's performance ultimately served to underscore the power held by the latter in the recording of his song, Altman's presentation of Tommy Brown singing in this live performance is spread across several images. Even though Brown is the star performer here, the total effect of these shot compositions as the song is performed do not suggest he possesses any particular power over this audience or its attention as Haven ostensibly commands in the recording studio. Close-ups of Brown singing are quickly joined in the editing with longer shots of the crowd gazing on at the Grand Ole Opry, the shots of the crowd having an aleatory effect that emphasizes the uncertain reception of the performance. In this crowd are also glimpsed a few characters from elsewhere in the film—L. A. Joan (Shelley Duvall), Private First Class Glenn Kelly (Scott Glenn), and the eventual assassin of Barbara Jean, the aforementioned Kenny—but only Joan seems to be particularly enjoying the music. Altman's strategies in assembling this array of shot compositions, all of them busy with performative detail, is made more complex by a pair of cutaways, during Tommy Brown's song, to shots of Albuquerque (Barbara Harris), an aspiring singer who yearns for a position of celebrity very like the one Brown currently enjoys. Altman's match cuts to these shots of Albuquerque deftly match Brown's position in the widescreen frame to Albuquerque's, a gesture that underscores her desire to occupy Tommy's position in the wide frame even as it also suggests his own relatively marginalized position in that celebrity world (he is, after all, performing here as an opening act to Haven and Barbara Jean, implied by the film to be on higher rungs in this fictional Nashville's celebrity hierarchy). These images and their array of potential implications are nevertheless driven by Tommy's rollicking song, generating a rhythm and apparent unity. Altman's assemblage of the sequence through these various shots ultimately points to the complexity of reception, in this sequence in which Tommy has the attention of a large group of people even as the effect of his performance and his celebrity are uncertain.

This idea that the powerful stars in the world of *Nashville*, because of or perhaps even despite their power, are always subject to the uncertain reception of their audience is most bitterly and acutely conveyed by the complex arrangement of images in the film's final musical performances, presented in the sequence in which Barbara Jean is assassinated while performing onstage for a political rally for the film's fictional presidential candidate. The sequence begins with a shot of the American flag, which takes up the entirety of the Panavision frame, as an announcer on a loudspeaker introduces Barbara Jean and Haven Hamilton to applause. As we hear the applause fade out and the music begin, the camera zooms out and pans down to reveal Barbara Jean and Haven Hamilton onstage, surrounded by musicians and with a political banner for Thomas Phillip Walker behind them. The initial wide frame in which the American flag constituted the entirety of the image, presumptively implying ideological unity in the people gathered to hear Barbara Jean and Haven sing, is immediately complicated, and contradicted, by Altman's aleatory cutaways to the various individuals present in the gathered crowd. As Altman's frames linger on these various people, space is created to observe that some of them seem interested in the performances onstage, while others bear the contingent reactions and distracted attention one might expect of an audience gathered to play "extras" in front of a fictional group of singers and musicians during the production of a movie. These shots also include images of some of the fictional characters from the film who we see inch their way to the front of the stage. These characters include the eventual assassin Kenny, to whose intense gaze Altman repeatedly returns, in conjunction with shots of Barbara Jean, who sings a solo after Haven leaves the stage. From among all these various gazes that are cast up at Barbara Jean, Altman creates a kind of terrible intimacy between the singer and Kenny through repeated cutaway shots to the latter. The extras surrounding Kenny in the frame, some of whom glance at him with a casual suspicion, seem to sense that all is not right with him. In the last of these cutaways, Kenny looks up and offscreen. The subsequent shot, implied to be the object of Kenny's gaze, is a second wide image of the American flag, again engulfing the Panavision frame. We never return to a subsequent reaction shot of Kenny after this image of the flag and before the assassination. Barbara Jean finishes her song. Haven comes out with a bundle of flowers for her. And in a long shot that keeps some of the crowd and the entirety of the stage in the laterally arranged field of vision, a gunshot rings out. Barbara Jean falls.

The effect of this string of shots is to situate the assassination of Barbara Jean as something apart from any particular psychology in this vast crowd, despite the fact that Altman's repeated cutaways to Kenny underscore his individual responsibility for the act. The terrible creation of this assassination, Altman indicates through the combination of these various wide frames, is as collaborative as the making of any one of the songs we have heard in the film, even if the reasons for it and responses to it are as ambiguous as the reception of the songs themselves have been throughout the movie. But Altman also eschews articulating what kind of "unity" creates and then responds to such acts of violence, leaving any judgment about that to the viewer. Despite his terrible act in this sequence, Kenny occupies no visual site of power in the wide image in which Barbara Jean is killed. The emphasis in that long shot is on the effects of his decision to fire his gun: Barbara Jean falling on the stage, some in the crowd screaming. But it is nevertheless still a long, wide shot depicting a crowd and not an individual act. In a subsequent and very busy shot of the crowd, Scott Glenn's Private Kelly, the lovestruck admirer of Barbara Jean who has been following her throughout the film, overtakes the assassin. But for Altman, the heroism of this act is immediately diffused by cutaways to surrounding, aleatory chaos, in the crowd and on the stage. In a frenzied attempt to restore the apparent "unity" ostensibly symbolized in the two shots in which the American flag has momentarily commanded the entirety of the Panavision frame, and the event of this political rally itself, Albuquerque, who has been lingering in the wings of the stage throughout the performance, is given the microphone. She begins singing a rendition of "It Don't Worry Me." Altman includes more cutaways of the crowd but nevertheless gives Albuquerque pride of place through medium shots of relatively long duration as well. Some of the crowd are singing along with her. Others, who apparently have not yet managed to forget the fact that a person was just murdered in front of them, are not. The film ends on a slow pan up to the clouds, the ideological unity toward which the earlier shots of the American flag gestured now replaced by the ether of the blue sky. Any particular response to these images in *Nashville* is deftly anticipated and situated by Altman as part of an uncertain field of reception. In a country this big and in a frame this wide, the effects of celebrity image-making and politics are at best uncertain, at worst disastrous.

Short Cuts

Several critics perceived the mosaic film *Short Cuts* as part of Robert Altman's supposed "comeback" in the early 1990s, after the critical success of *Vincent & Theo* (1990) and *The Player* (1992). *Short Cuts,* Altman's only film with cinematographer Walt Lloyd (a film and television veteran whose screen credits also include Charles Burnett's 1990 film *To Sleep with Anger* and the 1990 teen drama *Pump Up the Volume*), is indeed a return to the kind of mosaic, multiprotagonist, widescreen filmmaking that defined Altman's film practice in the late 1970s. The film is a loose adaptation of a set of Raymond Carver stories—and one poem by Carver in particular, "Lemonade." The individual stories of twenty-two characters (from nine different families) are interwoven across over three hours of screen time, although the looseness of the adaptation renders any association between the film and the source text an act of creative interpretation. Robert T. Self has observed that the film operates according to the narratological idea of "compositional motivation," through which the viewer must forge meaningful links between causally ambiguous strands of plot and character (*Robert Altman's Subliminal Reality* 257). Striking widescreen image compositions cannot often be read in *Short Cuts* during the initial moment of viewing; one is reminded of Altman's words, cited at the beginning of this chapter, that in his films "you have to look among all these things, decide what's important; but you won't necessarily know it at the time." And the film, while ostensibly a throwback to Altman's late-seventies mosaics, does not evince the same bitterness as *Nashville* and others of that earlier period even as it also makes similar gestures to political subject matter (in *Short Cuts,* this involves the spraying of possibly toxic pesticides across the city in which all these characters live). *Short Cuts,* in its tone, is perched precisely between the irony of much of Altman's late 1970s work and the more earnest, consolatory tone of some of his later films.

Let us throw a line into the wide pool that is *Short Cuts* and try to fish out a connection. In the first hour of the film, three fishermen carouse by a small lake on the outskirts of Los Angeles. One of them, Vern Miller (Huey Lewis), standing on the top of a small stretch of rocks, urinates into the lake below while the others occupy themselves offscreen. This image of male swagger initially seems in keeping with Altman's typical ironic wielding of the anamorphic frame. Even though the film never encourages its viewer to occupy a position of stridently moralistic judgment (in this *Short Cuts* is similar to *Nashville*), we are already prepared to generally dislike Vern and his fishing associates because of their piggish behavior toward the waitress Doreen (Lily Tomlin) during breakfast prior to their fishing trip. And this is an attitude that the shots of the men by the lake does nothing to assuage. Vern's swagger is both defined by the image, with its lateral stretch of wilderness apparently complementing his rugged, devil-may-care attitude in the sequence, and also comically and even slightly cruelly undercut by it, given that the narrative hitherto has established that Vern is a working-class suburbanite rather than an authentic cowboy. Altman's Panavision frame also knows something about the callousness of Vern's behavior atop this rock before he himself knows it: the camera zooms in and pans down to the water in which Vern is presently pissing to find, floating under the water and near the surface, the dead body of a young woman. Vern will notice this body after the camera does and call his friends over, the ensuing conversation about what to do with the body now occupying the fishermen. Their insensitivity toward women, expressed earlier in the film, here expands to become a broader lack of knowledge about how to handle the remnants of human life. None of them seem to know quite what to do with the body.

Later in the film, it is revealed that one of the other fishermen, gruff Gordon Johnson (Buck Henry), has taken pictures of the corpse, after his photos end up by accident in the hands of Honey (Lili Taylor). These crude photographs, glimpsed briefly in Altman's frame as Honey unwittingly gazes upon them, diminish even as they preserve in an image the human life the remains of which they reproduce photographically, and are in counterpoint to the way in which Altman himself films the corpse in his wide frame. Altman repeatedly returns to the image of the corpse floating in the water when the film resumes, for a stretch at the beginning of the second hour, its narrative thread with the fishermen. The camera grants to this deceased person a gift she is no longer able to receive, the attention of the entire expanse of the frame in the creation of a poetic image. The image is poetic because while the frame gives us an image mercifully untethered to the crude perspective of the fishermen who have discovered this body, it also in the very same instant refuses to frame its presentation of the body in ways that are moralistically or cognitively superior to the fishermen's perspective, despite the relative omniscience of Altman's camera and despite the viewer's natural inclination to want a position of moral distance from these men. In fact, Altman brings irony even to his own ostensible omniscience. For, like these men and whether we like it or not, we do not know what to interpretively do, at this point in our viewing of *Short Cuts*, with this body. The wide frame acts as a magnifying glass drawing our attention to a life that has been extinguished and apparently ignored by everyone in this film's Los Angeles prior to this moment. The wide frame gives this deceased fictional person a place of visual importance that seems to substitute for the lack of characterological depth bequeathed on her. But the fact of our attention, guided as it is by the generous focusing of the widescreen frame on an unnamed figure who has no psychological bearing on the film's plot, is not enough to grant her honor. In terms of everything else we see in the immediately surrounding images of *Short Cuts*, and given that the behavior of the men has not tutored us in any kind of response we would likely want to be aligned with, we do not yet know what to do with her.

One recourse would be to situate this image of a dead body within a wider visual theme of death in *Short Cuts* and then see how this motif plays out with the various other images in the film. The son of Ann Finnigan (Andie MacDowell) and newscaster Howard Finnigan (Bruce Davison), whose editorials warn of the toxicity of the insecticide coating the cityscape of Los Angeles, will be hit by a car driven by Tomlin's Doreen and will, in the third hour of the film, die. A kind of contrapuntal poetry can be found in *Short Cuts* whereby the images of the dying boy are mentally juxtaposed, during a viewing of the movie, with the images of the floating corpse in the river. Where the corpse in the lake is depicted in Altman's wide frame alone, frequently from above in an ostensible position of grieving but also, given that we know nothing about this dead woman, abstract omniscience, the senseless and sudden death of a young boy, whom we and the parents thought was recovering in a hospital, is captured in a series of rushed images that glimpse his final breaths through the awful enclosure of internal frames formed by the bodies and hands of nurses and doctors trying to save his life. But Altman does not return to images of the dying boy for poetic effect in the way that he will return throughout the middle stretch of the film to the floating body in the river. This is perhaps because the passing of the young boy is met by overwhelming floods of grief from his parents, who have attended to every minute of his coma in the film's hospital scenes. The generous expanse of Altman's wide frame does not need to compensate emotionally or substantially for the boy's passing in the same way that it acknowledges the body of the more cruelly abandoned, anonymous young woman who does not receive a similar outpouring of grief from anyone in the film.

The fact of the dead young woman's nudity, too, and the film's compulsive way of returning to this dead figure lying nude in the lake could be poetically likened to nudity elsewhere situated in the film's wide images, particularly the nudity present in a scene involving a pair of sisters played by Julianne Moore and Madeleine Stowe. Just as the nudity of the dead young woman seems to serve some larger aesthetic purpose in Altman's frequent returning to it as an image worthy of attention, so is the nudity of Stowe linked to art, given the occupation of the Moore character. Painter Marian Wyman (Moore), wife of the doctor Ralph Wyman (Matthew Modine) whom we have seen attend to the ailing Finnigan son in the hospital scenes, paints slightly abstract and expressionistic renderings of the human figure. One of her subjects is her sister, Sherri Shepard (Stowe). In the second hour of the film Marian paints Sherri, posing in the nude, as Ralph, not knowing a session is in progress, walks in. Roughly similar to the fishermen in their discovery of the naked corpse but now faced with the added complication of a living interiority behind the flesh, Ralph does not know how to respond to the fact and presence of Sherri's bared body, and after he bumbles out of the room, the sisters laugh derisively at him. Marian's pictures, or at least the ones

FIGURE 3.5 *Short Cuts* (Fine Line Features, 1993). Digital frame enlargement.

we are privileged to see in the frames of *Short Cuts*, often capture figures in the act of laughing, their expressions in her paintings often closed off in a frame of consumed self-satisfaction. In this way these paintings, although decorating the lateral stretch of Altman's mise-en-scène in the scenes set in Marian's house, are themselves isolated from the larger canvas of *Short Cuts* and its own aesthetic sensibility. This is a film that never quite encourages a tone of derisive laughter toward its characters, even as it shows that attitude existing within its world.

A few of Marian's paintings can be seen arranged around Sherri as she poses for Marian (Figure 3.5). She is positioned in the middle of the image on her stool as Marian's images of laughing and grinning figures internally frame her. The expression on Madeleine Stowe's face—eyes closed, knowing smile, face leaning upward as if away from any interlocutor who might disturb her playful confidence—rhymes with the attitudes of the figures in the paintings, the canvases of which surround her in the frame. The expression of mocking laughter is also in keeping with what we know of Sherri up to this point in the movie, of the way she responds with comic disdain to her husband, the cop Gene Shepard (Tim Robbins), when he makes up absurd stories about his profession as an undercover cop meant as covers for his late-night philandering. At the same time, the fact of Stowe's nudity is also thrown into relief against the workaday artistic studio in which she poses, and against the fact that the other figures surrounding her are painted in close-up. The composition of the shot has a similar aesthetic effect as, presumably, Marian's paintings do in the world of the film (elsewhere mounted on Marian's walls, out of the view of this shot, we have seen other nude portraits that combine cackling facial expressions with nudity). A counterpoint is developed here between the nude image of Stowe and the poetic image of the dead body elsewhere in the film. Stowe's nudity, like the defiant expressions of refusal inscribed on the figures in Marian's paintings that surround her and inflect her own performance as a model for Marian, is part of an aesthetic, social, and personal context of meaning that the dead, anonymous woman does not enjoy and which, given her death, is now impossible for her to become a part of, despite her ostensible "participation" in a social network of visual and narrative complications in the film that is *Short Cuts*. Stowe's nudity is presented by the film, without prurience or prudery, as an artistic fact, something to be preserved in pigment on canvas by Marian, whose artistic studio and interpersonal relationships define the mise-en-scène of this frame and in ways that throw into relief the now permanent lack of such situated, social definition for the floating corpse, even though both women occupy the same anamorphic frame whose laterality enables social inscription in an unfolding tapestry.

We never quite get a sense of the reception of Marian's paintings in the world of the film, however. The irony of the wide frame of *Short Cuts* is that the film places these characters in an expansive social network without giving us any substantive sense that the labor they perform, aesthetic or otherwise, means much of anything to the other characters. Marian's husband, the caustic doctor played by Modine, does not appreciate her artworks; he indeed wonders why nudity is enough to bequeath upon her paintings the aura of "art." His words are a wry questioning of the film's own use of the nude female figure as a site of serious, adult attention in the wide frame: Why focus on such relatively trivial aesthetic matters when other frames in *Short Cuts* have opened up a world of other problems, including environmental, economic, and interpersonal ones? Nevertheless, the doctor's generally sarcastic and apparently ignorant attitude toward art and art-making does not make him an authority on the subject, and certainly not in relation to the more generous perspective of the film itself. Although it is implied that Marian has enjoyed some success as an artist, *Short Cuts*, unlike the presentation of the country songs in *Nashville*, does not give us much space to think about the reception of her art beyond the doctor's comments and beyond our responses to what we see of her work. And our response to her art will be as diverse as the various assessments of the songs in *Nashville*. But unlike the songs in *Nashville*, which seem to at least carry the potential of larger cultural meaning, the paintings in *Short Cuts* are personal works located in a socially circumscribed milieu. The omniscience of *Short Cuts* is sensed not only in the film's narrating, from a relatively detached position, the experiences of twenty-two characters. This omniscience is also generated by its image-making, its ability to visually situate characters in a wider anamorphic frame that repeatedly outstrips any individual perspective even as the frames and the performances within them vividly inscribe divergent subjectivities. This is typically a gentle rather than a masterful or cruel visual omniscience in *Short Cuts*, one that bends toward understanding and consolation more than it does toward causticity and irony. Marian's paintings, in their angularity and obsessive focus on cackling faces, are perhaps ripe for parody. But Altman's frame opens up to other perspectives, and we are in turn distanced from our own potential jeering of Marian's work by her husband's rather daffy regard for art in general, a position of ignorance with which, like the point of view of the insensitive fishermen, we would not wish to be aligned. The wide frame in *Short Cuts* offers us only apparent mastery of a visual world via its anamorphic frame but undercuts us repeatedly, not to show off its own knowledge or to position us, or the characters, as stupidly ignorant but rather to remind us of something we might have missed. The visual omniscience of *Short Cuts*—both the totalizing effect of the film itself and the quality of observation present in any one of its frames—has a power to create a full, evocative, enthralling image, but this is nevertheless a modest power that Altman wields carefully, and one that in his hands is attentive to human limits as much as it is to human connection.

From out of this gentle omniscience, which places the relatively sharp irony and lament of the earlier *Nashville* itself at arm's length, *Short Cuts* marks a first gesture toward a tone of consolation in Altman's late work, films in which artistic performance is situated not as a crass stab at commercial celebrity or smug ideological calculation (as it was in the milieu of *Nashville*) but rather as an earnest and potentially fulfilling way of coping with life. This relative optimism is evident in a work made after *Short Cuts*—Altman's penultimate film, *The Company*.

The Company

Here I take a brief sojourn to a film with an even more optimistic sense of the consolation art offers and which Altman makes ten years later, before returning to *Short Cuts* as a means of bringing to a close this chapter's discussion of Altman's anamorphic cinema. *The Company* tells a story of dancers striving to succeed in the Joffrey Ballet in Chicago. Three of the film's many characters are fictional: Ry (Neve Campbell), one of the company's top dancers; Josh (James Franco), the cook who becomes, halfway through the film, Ry's lover; and Mr. A (Malcolm McDowell), the testy but ultimately endearing company director. Much of the rest of the cast is made up of the talented dancers of the Joffrey company, presumably playing versions of themselves. Campbell's involvement with the project enhances the documentary charge in this fiction; Campbell was a dancer before becoming an actress, and the film originated with her and screenwriter Barbara Turner, who brought the project to Altman. The film marks another in the many uses of the 2.35:1 frame in Altman's career, but particular technical aspects *The Company* employs are nevertheless distinguished. As Robert T. Self has discussed in his essay on the film, in filming the ballet performances by members of the Joffrey company, "Four separate camera set-ups covered the stage, and each camera fed its own monitor in a bank of monitors arrayed behind the stage-left curtain" ("Art and Performance" 161). Using these multiple, high-definition digital cameras in concert with cinematographer Andrew Dunn (the first use of digital by Altman in his career, which he would repeat in his use of the Sony CineAlta HDW-F900 digital camera in the filming of his next and final film, *A Prairie Home Companion*), Altman was able to review footage immediately and make adjustments to subsequent takes on the set as necessary, an important resource in the staging of complex dances by ballet dancers who risk injury if pressed to perform multiple takes. The multiple camera angles created by this quartet of cameras not only provided Altman with varieties of coverage that could be used in assembling the final cut, but were also suitable for the subject. The film itself, as both Altman and Self have suggested, dances around its subject matter, nimbly moving from one point to another (both in the filming of the dances and in the behind-the-scenes interactions in the company). This filmic dance is all the more impressive given that the frame shape that is moving so nimbly, through the use of these lightweight visual cameras, is in 2.35:1, a ratio historically and popularly associated with bloated spectaculars. What results is a film with a more optimistic attitude toward the creation of art, in this case dance, than earlier Altman films. Although the film makes clear to us the fragility of the ballet dancer's body, there is no ambiguity expressed in *The Company* over the reception of ballet as an artistic practice. The admiration for the dancers is expressed by the film's own eloquent, balletic movements in filming them, an appreciation that is presumed to find its equivalencies in the audiences gathered to watch the dancers of the Joffrey perform in the world of the film.

The Company has many of the trappings of the conventional Hollywood dance picture, including a behind-the-scenes romance between the aspiring dancer Ry and her boyfriend Josh, who has displaced a former lover hoping to win Ry back. But Altman defuses this romantic ménage à trois before it even begins, relegating Josh largely to the background (often literally in the background, as a viewer of Ry's performances) as the film proceeds to showcase its interest in ballet dancing. So too does the narrative cue expectations that much of the story will be about Ry's successful attainment of a winning position in the Joffrey dance troupe. But Ry's (and Campbell's) most eloquent and convincing dance comes early in the film, during a courageous performance in an outdoor amphitheater. A thunderstorm begins midway through the performance, and Ry must dance on an increasingly slippery stage. Altman's camera shuttles between Campbell's beautiful dance, with audience members in the foreground of the frame opening up umbrellas to protect themselves from rain, and shots of Mr. A and others in the company, initially seated in the audience before rushing to the wings of the stage, hurriedly trying to ensure that the conditions are presently safe enough for Ry's performance to proceed. Altman's dance with the camera here involves not any kind of ambiguity over the public reception of the performance, which given Ry's astonishing feat in the rain is bound to be positive. The wide frame in *The Company*, rather than opening up a space for a condescending response to earnest art (at least one possible position enabled by the wide frame of *Nashville* and perhaps even by the relatively empathetic wide frame of *Short Cuts*), instead gathers up into its view a community of artists and art appreciators who express great care and love for the art they are witnessing. This expansiveness of feeling and generosity is felt throughout the film (although it is a feeling occasionally

FIGURE 3.6 *The Company* (Sony Pictures Classics, 2003). Digital frame enlargement.

punctured, in a way befitting the subject matter, by the competitive nature of the world of ballet dance and by the inevitable disappointments in the careers of most dancers) (Figure 3.6).

The ending of the film indeed reaches beyond its own narrative machinations to embrace other experiences, as if in response to a lament by Mr. A, in an earlier scene, for the gay dancers in the company whose lives were cut short by AIDS. Bucking convention, the film does not end with Ry's triumph in the final dance. Instead, Ry injures herself during the performance and must be replaced with another dancer. *The Company* displaces its own star, Neve Campbell, enabling the presence of the dancers of the Joffrey company to command the expanse of the frame, from all of its multiple views in the auditorium. A similar displacement can be found in the culmination of the romance between Ry and Josh. Josh has arrived at the final performance with flowers for Ry. The final romantic clinch between them does occur, but Altman displaces it in his anamorphic frame. As already mentioned, Ry, before Josh's arrival, has injured herself. She is not a part of this dance. When Josh arrives with his flowers, he is on the wings of the stage, but Ry, her arm in a cast, is on the opposite side of the stage. In order to get the flowers to Ry, Josh must somehow dart across the stage (potentially interrupting the ballet) (Figure 3.7). He does so during the curtain call ("Who is this guy?" one puzzled dancer asks), his movements from right to left marking a thrilling bound across the lateral stretch of the frame but one that Altman inflects with cuts that bring us broader views of the stage, the other dancers, and the cheering crowd. This ending satisfies the expectation that the romance will be resolved but also situates it in the wide image as only one part of the artistic life the film celebrates.

The relationship between aspect ratio and subject matter in *The Company* ultimately works to place artistic achievements of highly talented individuals in a larger, communal frame. Even when individual dancers are isolated or atomized

FIGURE 3.7 *The Company* (Sony Pictures Classics, 2003). Digital frame enlargement.

at certain points in the film—after, for example, one dancer is replaced by another during a rehearsal that has gone awry, or after one dancer tragically snaps her tendon (effectively ending her career)—such experiences of acute misfortune are expressed by *The Company* as an unavoidable part of the realization of any artistic achievement, necessary tragedies that engulf particular lives but which, alas, must occasionally happen for anything as remarkable an achievement as a marvelous ballet production to occur at all. These feelings of consolation in the face of personal failure, including the bittersweet nature of Ry's failure to single herself out as a star as expected in the final dance, ultimately situates *The Company* as one of Altman's more optimistic films. It is a work in which the widescreen frame distances us not to place us in an ostensibly superior perch of knowing but to glimpse the warmth and satisfaction that can be found in the collective, competitive, and ultimately collaborative achievement.

Coda: A Return to *Short Cuts*

To return to *Short Cuts* as a means to bring this chapter's discussion of Altman's widescreen cinema to a conclusion, we can observe that these collective consolations in the achievement of art on display in *The Company* are precisely what Zoe Trainer (Lori Singer) is missing. Zoe, an accomplished but, the film eventually makes clear, psychologically tortured cellist, is repeatedly situated by Altman's frame as an atomized presence. The viewer's interpretive struggle to meaningfully place Zoe alongside the other characters in *Short Cuts* is seemingly shared by the film itself, which initially places her at a distance. She is first glimpsed in the opening credits montage, which introduces all the characters, during a performance in which her cello is accompanied by four violinists. The camera, initially presenting her as one, albeit central, musician among others, begins to zoom in gradually toward her, underscoring her eventual centrality as a character later in the movie but also atomizing her from the other musicians. The

FIGURE 3.8 *Short Cuts* (Fine Line Features, 1993). Digital frame enlargement.

warmth and fixed attention that will greet the Joffrey dancers in *The Company* are not gifts bequeathed to Zoe. Altman's cutaway to a shot of audience members we will also soon become acquainted with as characters in the film (those played by Julianne Moore, Anne Archer, Fred Ward, and Matthew Modine) draws our attention to the distracted spectatorship of Zoe's audience (these four characters are chattering away, making weekend plans, rather than listening to Zoe's music). That these characters, although intertwined in the same narrative structure as Zoe, do not in fact know the cellist at all only serves to retrospectively underscore her isolation during this performance (even as Zoe's music continues to "score" the disruptive conversation between these characters as they ignore her music).

This isolation of Zoe's performance in the wider anamorphic frame is a repeated motif throughout the film. When she next appears, the sound of her cello wafts from outside her bedroom, in a shot that initially focuses on Jerry Kaiser (Chris Penn), who has arrived to clean the pool of the Finnigans, Zoe's neighbors. Altman's camera zooms in toward Zoe, bringing us at once closer to her performance of the cello but again isolating her in the frame, this time in the internal frame created by the window through which we glimpse Zoe's performance from the outside (Figure 3.8). The effect here, as it was in the earlier performance, is to situate Zoe as the creator of a soundtrack to a narrative experience in which she has no meaningful part. The sound of her cello embraces the expanse of the anamorphic frame even as that frame itself underscores the way in which Zoe is isolated from others. However, Zoe will attempt to become a part of this narrative flow. She is later shown playing basketball with a group of men in the driveway to her apartment complex. As she plays, Casey Finnigan, not yet the victim of the impending car accident, walks by. Zoe tries to connect with him, asking him if he wants to watch her dribble the basketball. The moment suggests a possible bond between Zoe and her younger neighbor, but also Casey's own isolation as he strolls past without acknowledging her. The frame, which again slowly zooms in toward Zoe after it presents the failed interaction between her and Casey in a wider frame, allows us to view Zoe's attempt to connect even as it once again isolates her as the figure of momentary interest in the frame. It is also the first time in the film that Zoe becomes associated with the theme of death, in her attempt at connection with the doomed Casey, although this idea emerges only in retrospect.

Zoe will commit suicide near the end of the film, poisoning herself with carbon monoxide by running her car in a locked garage. That Zoe and her presence in Altman's wide frame can only be interpretively fulfilled after we know she will die gives every encounter with her a melancholic tinge on repeat viewings of *Short Cuts*. Her placement in the frame repeatedly prefigures her grim end but without the viewer, on a first pass through the film, being aware of this finality. Altman's structure generates empathy and understanding for Zoe's situation even as it is complicit with the placement of Zoe in a narrative that must rely on her suicide for thematic resonance. Like Sherri and Marian, Zoe is also glimpsed in the nude but not for any aesthetic effect and not as the object of a painter's gaze. Her baring of her body, which is glimpsed by Jerry, during a break from his pool work, through the interstices of a wooden fence as he speaks on his mobile phone with his friend Bill Bush (Robert Downey Jr.), is presented in a voyeuristic way and, again retrospectively within the tapestry of *Short Cuts*, as part of Jerry's own troubled relationships with women. She dives into the pool at which point Altman cuts from Jerry's detached perspective to an overhead shot of Lori Singer floating, facedown, in the pool. Whether she is playing at or seriously attempting suicide is at this moment in the movie disturbingly unclear (her basketball from the earlier scene here floats in the pool alongside her unmoving, floating body, a nod to the potential that this is a moment of play even as it might very well be a moment in which Zoe, keeping her thoughts and feelings from view, is truly ending things). All this forms an emotional undercurrent on repeat viewings when it becomes clear that this image is meant to rhyme with the eerily similar image of the anonymous dead body discovered by the fishermen in the lake. Failing to command the attention of others in Altman's frame, either through her music or through her other gestures of generosity, Zoe occupies a posture in the water that rhymes with other images of death in *Short Cuts* and that forecloses any connection with the wider world through her art, her music.

The moment is also a reminder of the kinds of omniscience that Altman brings to his widescreen cinema. Altman deploys irony as an aesthetic resource in his creation of widescreen films, using the anamorphic frame to pose to his viewers questions about what the achievements and existence of his characters might mean in a frame that repeatedly insists on a context largely indifferent to them. His wide frames open up a space of improvisation and discovery for his performers, who are given space and time to perform diverse facets of personality even as a certain malaise inflects the overall framework. In *Short Cuts* it is not always clear what these lives can ever amount to, given the complex nature of the film as an unfurling tapestry the entire view of which is a constantly shifting mental and visual construct. Although Altman's anamorphic cinema might ultimately be read as the product of a happy artistic collaborations of the kind on view in *The Company*, the director is also keenly aware of lives that do not find such fulfillment, whose achievements meet with failure in a wider social frame. We have seen such moments of failure throughout this chapter, in the form of the deaths of McCabe; Barbara Jean in *Nashville*; and Zoe Trainer and the anonymous dead body in *Short Cuts*. Altman's frame welcomes the open participation of the individual (the actor in his frame and the viewer positioned to make sense of the films) but also situates that participation in a film world—and Altman liked to always think of his entire body of work as one long film—that goes beyond, again and again and even as forms of consolation become possible in the later work, the ability of any one person to make peace with it.

4

John Carpenter (1948–)

Anamorphic Haunting

> On television, you see squares. And that's fine for television.
> —John Carpenter (qtd. in Appelbaum 11)

> Films discussed in this chapter: *Assault on Precinct 13* (1976); *Halloween* (1978); *The Fog* (1980); *The Thing* (1982); *Christine* (1983); *Prince of Darkness* (1987); *They Live* (1988); *Memoirs of an Invisible Man* (1992); *In the Mouth of Madness* (1994).

Murderous Prologue: *Christine*

The revving purr of a car's engine soon grows into a growl. Suddenly, the sound fades out. The first image of John Carpenter's *Christine*, shot by cinematographer Donald M. Morgan (who will work again with Carpenter the next year on *Starman*), by means of a panning Panavision camera swooping downward, drops the viewer into an automobile factory. Here, among others of its kind painted in white, is built a sleek, red 1958 Plymouth Fury. After its downward movement to the factory floor is complete, the camera moves a little more slowly. Plymouth vehicles unfurl along the assembly line, toward the camera. Fury after Fury, painted in glossy white, emerges along the depth of the shot's composition as workers inspect the automobiles. A cut to the next shot, with the camera now laterally tracking along the assembly line, shows us the steel, chrome, and rubber of the bottom half of the cars but also a pit beneath the cars from within which men, their faces illuminated by yellow light, inspect the machines.

And then the car to be christened Christine arrives. When she does—its eventual owner will insist she's most definitely a she—the camera reverses its rightward tracking movement, now heading leftward to follow the red car along the assembly line. The sleek, red Christine now commands the entire stretch of attention in Carpenter's anamorphic frame, its rich, vibrant color thrown into relief against the white vehicles surrounding her. In this way Christine is quite unlike the other monsters and bogeymen in Carpenter's films. Those terrifying figures inhabit the margins of the director's images, at times offscreen, waiting for the perfect opportunity to violently intervene. The murderous Shape, in Carpenter's original *Halloween*, for example, is most haunting when he goes unseen. But Christine is up to something else. Her form of terror takes on, from the get-go, a palpable presence and visibility in the frame. She makes her murderous personality immediately known in the wide image.

Automobiles and cinema, and widescreen cinema in particular, had a long history when John Carpenter made *Christine* in 1983. As Kathrina Glitre discusses in her study of the spectacle of consumerist consumption in 1950s widescreen films, automobiles typified the glossy style of many movies of the era, "with their two-tone colours, shiny chrome bumpers and tail fins," and "they also provided an ideal object to fill the widescreen—long, thin, spectacular" (136). Glitre subsequently notes that automobiles were also gendered in such films, the word "chassis" frequently used by film critics to describe not only the automobiles appearing in CinemaScope frames but also the bodies of stars such as Marilyn Monroe and Jayne Mansfield (136). Carpenter acknowledges this fetishism in his own widescreen Panavision frames in *Christine*, which involves a young man's obsession with a car. But Carpenter's car does not remain simply a sleek vehicle to-be-looked-at in the widescreen frame, and her antagonistic relationship to other women in the movie does not suggest any easy parallels between feminine automation and actual human beings. This car takes on a kind of singularly devilish agency as the film progresses, becoming the most vivid personality in the film.

Carpenter, in an implicit critique of the consumerist display of the automobile in earlier widescreen cinema—think of the cars in Clifton Webb's factory in Jean Negulesco's *Woman's World*—also shows how Christine and her murderous subjectivity are produced. In this 1957 factory in which Christine and her Plymouth chums are being built, everything grinds along efficiently, as the workers, equipped with their various tools, move busily across the frame, in apparent ignorance of the red monster they are unwittingly complicit in manufacturing. Carpenter cuts from this shot following the car to another leftward tracking shot, a fetishistic close-up of Christine's rear lines, as the car moves down the assembly line. The mobility and length of the anamorphic frame here are fascinatingly

FIGURE 4.1 *Christine* (Columbia Pictures, 1983). Digital frame enlargement.

conjoined with Christine's own sleek, glacial, chromic angularity. But this imagery is not simply the consumer's gaze. The camera spies something in Christine that goes beyond object-to-be-bought status. She is a product in revolt. Factory lights playfully reflect off of Christine's steel surface as both camera and car continue their journey down the assembly—it is almost as if the camera, guided along by Christine's glossy red, can already spy a cinematic spirit in Christine, as if Carpenter's Panavision lens were recognizing the potential cinematic qualities lying dormant in automotive technology. Another cut takes us to the front of the car and to another close-up: a rearview mirror in which is glimpsed internally one of the factory's workers following along from behind (Figure 4.1). He soon calls for the assembly line to stop—there is something in Christine's engine he would like to check. As he does, Carpenter's camera moves in a herky-jerky forward tracking shot in counterpoint to the smooth and efficient movement—of both camera and the onscreen work of these factory figures—defining the scene hitherto. Then a cut to a close-up, shot from an imposing low angle, of Christine slamming her hood down on the worker's hand; the frame, at first sensing the percolation of automotive personality, now relinquishes its detached perception in its own complicity with the terror Christine creates, as a cut to the next shot is timed with Christine's "bite" of the now-screaming worker's hand.

Christine begins by revealing a context—the industrial manufacture of the automobile—before Carpenter shifts focus to the car's incipiently murderous subjectivity. In Stephen King's book, the evil of this car is explained in fulsome backstory; rather than being "born bad" as a product of industry as Christine appears to be in this opening sequence in the Carpenter, the killer car in the novel is haunted by the malevolent spirit of its first, murderous owner (see Howarth 136 for more discussion of the adaptation). Christine's malevolence, in King's version, is the product of a human narrative: she is not evil until she is imbued with the psychology of one particularly malevolent individual. Carpenter dispenses with psychological explanation almost entirely. The scene in which the awkward teenager Arne (Keith Gordon) finds, twenty years later, a junked, red Plymouth Fury for sale will make some allusions to the difficult life of Christine's first owner. But Carpenter's camera ultimately sees the car's evil as always already existing in—as being produced by—the factory in which these workers dutifully assemble these sleek, devilish Plymouths. (Benjamin Stoloff's 1932 pre-Code film *The Devil Is Driving*, which also suggests an incipient murderousness in the modern automobile, is an important Academy aperture precedent for Carpenter's widescreen film.) Carpenter, and in ways that befit the director who will make even more pointed socioeconomic observation in *They Live*, sees this car's villainy as the product of the industrial society that creates it, rather than as the result of one isolated evildoer's actions. And his widescreen frame, itself an instrument created by a commercial industry, is obliquely complicit in these machinations, in its "technique of terror" (Conrich and Woods 1). Indeed, each of Carpenter's images is as sleek and as elegant and as puckishly haunting as Christine herself. But Carpenter distances himself from simple indulgence in genre conceits through his singular deployment of the widescreen frame, a technology that in his hands finds aesthetic harmony in presentation of unsettling subject matter.

As with the other three directors studied in this book, choice of aspect ratio is never only a vehicle for narrative information for Carpenter but rather a screen shape that develops complex relationships with subject matter. In Carpenter's films, these complex relationships between frame and subject are shaped in part by an attitude toward genre—in particular, horror and science fiction—that eschews the irony characteristic of Carpenter's contemporaries. In a 1999 appreciation, critic Kent Jones referred to Carpenter as "the last genre filmmaker in America. There is no one else left who does what he does—not Hill, not Cronenberg, not De Palma, not Ferrara, not Dahl, not even Craven, all of whom pass through their respective genres with ulterior motives or as specialty acts, treating those genres as netherworlds to be escaped to, museums ready to be plundered" (26). Carpenter, for Jones, rather works with an artistry "focused on satisfying genre conventions and the demands of narrative," with his "loftier preoccupations" realized through genres rather than through condescension toward them. For Jones, Carpenter is also

> the widescreen master of contemporary cinema. . . . Along with Minnelli in his Fifties melodramas and the Resnais of *Last Year at Marienbad*, Carpenter is one of the only filmmakers who bring the shape to life, just as the 1.85 aspect ratio becomes a living entity in Spielberg's work and 1.33 does in Murnau and Lang. The Scope frame is often associated with deserts and windswept vistas, and touristic epic sweep. Not to deny David Lean his place in history, but in comparison to Carpenter his "immaculate craftsmanship" is alienated and plodding—Alma Tadema to Carpenter's Homer. (28)

As Jones suggests, Carpenter's films satisfy the conventions of genre. Like Christine herself, her sleek lines custom-made for Carpenter's compositions, the Panavision frames of the director's films are carefully crafted images that serve the assemblage that is story. But his wide images go beyond this functionalism. For Carpenter, the wide frame has the potential to haunt his viewer, his images lingering in their resonance beyond the immediate moment of narrative consumption.

"The first time I used Panavision," Carpenter once remarked of the widescreen technology that defined the parameters of "Scope" after CinemaScope technology itself became obsolete, "I thought, 'This is like painting a picture. Look at the room you have, on the sides. You can *use* the space'" (qtd. in Fox 40). Carpenter's alien beings and horrifying creatures can use that space, too, can be placed, visibly or invisibly, in various surprising and suspenseful arrangements in the director's orchestration of the Panavision frame.

The Ghosts of Academy Aperture: *Assault on Precinct 13*, *Halloween*, and *The Thing*

Very early in his career, after the spectacular success of *Halloween*, John Carpenter made clear his taste for anamorphic imagery. When asked in 1979 why he chose the 2.35:1 Panavision format to stylize *Halloween*, he responded:

> I love Panavision as a composing rectangle. There seems to be two really good visual ways of composing. One is the old-fashioned format which they never use any more, 1.33 [*sic*]. It's a square. Beautiful to compose. And then there's Panavision, which is also beautiful to compose. It's perfect for the two shot. 1.85 [*sic*] is a bastard because you can't compose anything in it. It's not wide enough or tall enough. And there are other reasons. In projection, you're never going to get 1.85 like you have it; they're always going to break it up or down. And it really doesn't lend itself to making pretty pictures. You have a close-up in 1.85 and there's a little space off to the side. Well, in Panavision, you've got background and foreground objects to play with. I just love Panavision. It's a cinematic ratio, and I don't think you see it anywhere except the movie house. (qtd. in Appelbaum 11)

The anamorphic format enables Carpenter, as he puts it, to create "pretty pictures," a perhaps surprising phrase given the expectation of gruesome imagery in his chosen genres. But for Carpenter the format also evokes a specific public venue—the theatrical movie house—that at the time of the 1979 interview was still the primary location for encounters with the wide image. All but one of Carpenter's theatrically released films has been filmed and composed using various iterations of Panavision cameras and lenses, from the Panavision PSR R-200 on his first major theatrical feature, *Assault on Precinct 13*; to the Panavision

Panaflex (accompanied by the elegant, Steadicam-like moves of the Panavision Panaglide during handheld mobile framings) of the work spanning *Halloween* to *Starman* (1984); to later variations of Panavision in the late 1980s, including the Panavision Panaflex Gold used on *They Live* (and the Gold II on 1994's *In the Mouth of Madness* and 2001's *Ghosts of Mars*). The director's devotion to each new iteration of Panavision technology is not that of a technological fetishist (he is quite unlike Arnie in *Christine*, and not akin to technologically obsessed filmmakers such as James Cameron and George Lucas) but rather that of a director committed to a format best suited to his cinematic eye. Long after directors had turned to Panavision in order to mitigate (if not entirely eliminate) the optical hiccups that early CinemaScope introduced to moviemaking—for example, the "bending" of vertical lines near the edges of the frame—Carpenter would use the technology precisely to create the kind of optic materiality generated by its limitations, using Panavision in combination with extreme wide-angle lenses precisely to create curvature near the edges of the frame (see Figure 4.9, later in this chapter, from *In the Mouth of Madness*). Carpenter in this way approaches Panavision for its aesthetic properties rather than for its enabling of incremental technological "innovations" that increasingly repress, in favor of a kind of anodyne digital "smoothness" or glacial quality, the viewer's awareness of the unruly optical materiality of lenses and cameras.

As Sheldon Hall points out in his perceptive essay on the widescreen aesthetics of *Halloween*, Carpenter's affection for the Scope frame is perhaps surprising given that the filmmakers he most admires, in particular Howard Hawks, worked mostly during the era of cinema (the 1930s and 1940s) in which the rectangular widescreen format was not an available choice for shooting pictures (68). Hawks himself, whose film *The Thing from Another World* (1951) would form the basis of Carpenter's remake *The Thing* in 1982, was no fan of CinemaScope technology. Hawks used the format only once, on *Land of the Pharaohs* (1955), a striking and unusual epic that left its director dissatisfied. Carpenter is quoted as saying that his fantasy is to have made films during the classical Hollywood studio era of the 1940s (Hall 68), the same era in which Hawks worked. But despite Carpenter's nostalgia for an era of cinema in which his beloved Panavision was unavailable as a technical choice, we should remember, as Murray Pomerance reminds us, that CinemaScope pictures were not perceived by their first audiences as always necessarily *bigger* than Academy ratio (1.37:1) films, since the remaining picture palaces still lingering in many cities in the 1950s had very large screens regardless of aspect ratio, in other words in terms of height (see *Color It True* 2n). One could be overwhelmed by any number of cinematic shapes during the 1950s, the decade of Carpenter's childhood and the site of many of his formative cinematic experiences. And given Carpenter's stated appreciation for the compositional possibilities of 1.37:1 Academy aperture in the works of Hawks and others, it should not be difficult to imagine the director thriving in a technological and cinematic environment in which the "square" was the only available shape. It is only that, by the time Carpenter's career in cinema takes flight in the late seventies, his choices of frame for his theatrical features are 2.35:1 and what he calls the "bastard" shape, 1.85:1—which, for Carpenter, is really no choice at all.

Carpenter's initial pair of "siege" films, *Assault on Precinct 13* and *Halloween*, employ a poetic dialogue between the two frame shapes he finds especially attractive, 1.37:1 and 2.35:1, with the former taking on the form of an internal frame within the image and with the latter defining the overall shape of the film itself. These two films, like all of Carpenter's theatrical features from 1976 on, are shot in an aspect ratio of approximately 2.35:1. But both works also make pointed use not only of editing—that formal device that critics initially worried was the exclusive province of the 1.37:1 frame, and a limited resource in what was perceived to be the more theatrical CinemaScope—but also of aperture framing, the creation within the Scope frame, via some combination of vertical and horizontal lines within the wide image, of internal frames that evoke approximately the relatively square shape of earlier, pre-Scope aspect ratios. Academy aperture, in other words, haunts Carpenter's wide cinema from the very beginning.

Assault on Precinct 13 (shot by cinematographer Douglas Knapp) opens with a sequence depicting an assault not on but *by* police—in Anderson, California, "a Los Angeles ghetto," as a screen title tells us, in the dead of night. Their quarry are six members of an apparent gang, carrying weapons and strutting cockily. Although the diagonal movement of the gang members toward screen right indicates some kind of intention and purpose and eventually motivates, after the quick swivel of the camera, Carpenter's reframing of the gang from a low angle as they walk up a set of stairs in a very narrow alley, we do not as a result of this change in camera position receive any attendant sense that this group is gaining in strength or power in their uncertain purpose. Our not-quite-knowing what these people are up to as *Assault on Precinct 13* begins places restraint on the film's powers of narrative omniscience—important in a siege film eventually engaged with only a handful of characters. Despite the uncertainty that surrounds narrative events at this moment, the shot has a centripetal quality, drawing our attention toward a center (the center in the alley, marked at the end of the sequence by a pallid yellow light at the horizon) in a way that would not be illegible in the narrower confines of a 1.37:1 aspect ratio.

This first major sequence of Carpenter's career relies saliently on a technique initially thought to be unfit for widescreen cinema: montage. His masterful editing becomes apparent when the police, lingering from a position above the exterior walls of this dark alley, shoot down at the gang in the alley below. Carpenter's cuts to the police alternate between two shots, each filmed with a stable camera, in which the line of a shotgun, pointed down into the alley and firing away, forms a diagonal line stretching across the middle of the screen from either, alternately, upper left to bottom right or upper right to bottom left. This diagonal tension between the shots of the police positioned on either side of the alley, punctuated in between by shots of the gang members they robotically gun down, evokes Sergei Eisenstein's montage cinema, in which a picture of a whole event is generated not through an expanse of mise-en-scène but rather through juxtaposition of fragments. The diagonal line of the gun in each shot is matched with the diagonal line of the stone wall behind which the police linger, and the stable shots of the police firing their guns are put into counterpoint with cutaways to the victims in the gang, still presented by Carpenter with the same handheld camera movement. In all these shots the police brandishing guns are not identified as individuals; the top of the frame line figuratively beheads them, suggesting their status as instruments of state power. This use of social types rather than deeply individuated personalities in his presentation of the police and the gang also aligns Carpenter with Eisenstein, who famously used typage rather than substantive psychological articulations in the human figures in his films. This choice assuages—and, in the wake of the film to come, ultimately deflates—any parallel sensed between Carpenter's sudden announcement of his stylistic prowess and the power of the police. Carpenter's style has an aesthetic complexity contrapuntal to the frequently narrow-minded, self-involved fixations of authority figures in his films. This abstract presentation of the police in *Assault on Precinct 13*, across alternating, tightly composed diagonal shots, is especially notable given that there is ostensibly plenty of space in the wide frame through which Carpenter might have inscribed the personalities, or at least the faces, of these policemen.

Carpenter's stylistic choices complicate presumed understandings of even those characters in his films with whom we are apparently aligned. One could argue that Carpenter's refusal to fully identify with or explain the larger socioeconomic rationale of these "gang members" in the opening sequence is a sign that he is not a card-carrying progressive and that his film, it follows, is on the side of the police. Carpenter's composition of the anamorphic frame in relation to subject matter complicates any such judgment. The narrative, admittedly, develops psychological disidentification with the gang members who plot revenge on the police in retaliation for the deaths of the six gang members in the opening sequence. These vengeful figures are presented by Carpenter without sympathy; in fact, one of them, during an especially startling moment early in the film, is responsible for the death of a child. Yet the gang, taken as a whole, is in at least one sense very acutely aligned with Carpenter's frame and in ways much more interesting than the film's surface-level narrative content. This idea becomes visible in the gang's eventual ability, in the enacting of their ensuing, vengeful siege on the precinct, to learn how to move across the lateral stretch of the wide frame.

In the opening scene, as the gang is victimized by gunfire, their deaths are stylistically presented through jaunty handheld camera and montage cutting. As if haunted not only by the deaths of their brethren in the gang but also by the particular aesthetic shape in which those deaths have been presented, when the gang members later in the film begin their siege on the precinct, they do so in ways that play with, and at times invert, the stylistic logic of the murderous opening sequence. In the opening shots, the gang members are bathed in ominous shadow, and boxed in by the lines of the alley's walls extending toward the center of the frame, an aesthetic strategy that draws the viewer's attention inward, intentionally closing off a substantial portion of the frame. In the later siege sequences in *Assault on Precinct 13*, it is the lieutenant, Bishop (Austin Stoker), and the people gathered in the precinct under siege and whom he has the duty to protect (including the shell-shocked and grief-ridden father of the child murdered earlier in the film) who eventually become draped in darkness after the power lines to the precinct are cut. But most of these shots do not stylistically "box in" the characters in the precinct with aperture framing or internal frames. Carpenter repeatedly emphasizes in his orchestration of cutting in the sequences inside the precinct just how much interior space is at play here; when the film cuts to another camera position, another window, another door, another point of entry becomes visible, points through which a member of the gang might suddenly emerge. Further, where montage in the opening sequence is aligned with the diagonal authority of the "headless" policemen gunning down gang members across alternating cuts, in the siege sequences of *Assault on Precinct 13*, Carpenter uses the resources of montage editing to depict the interior of the police station as it is pelted with gunfire by gang members looming outside. The gang, in contrast to the constrained movements of their six fellows in the alley at the beginning of the film, are presented during the siege sequences, as they assemble themselves outside the police station leading up to and after various bursts of gunfire, as figures who now know how to use the lateral expanse of the frame. It is as if they have learned from the failures of their dead brethren. In one shot, two gang members move in synchronicity as they slip under a pair of trees, their balletic movements gracing their violent intentions with unsettling elegance. Later, after the initial attack of gunfire on the precinct, the gang must regroup for their attack's next phase. Carpenter frames the gang moving in diagonal cross-currents in a parking lot across the street from the precinct, their diagonal cutting-across of the wide image a surprisingly elegant figural appropriation of the similarly diagonal, but immobile and impersonal, power of the police in shots from the film's opening sequence. Here, as in the earlier shot of the pair slipping under the trees, the gang members come off less like a band of villains and more like a modern dance troupe who take a parking lot and a line of trees as their stage. In the way they upturn the film's earlier stylistic associations between film technique and character, and in the way they inhabit with such surprising elegance the lateral stretch of the frame with their synchronized movements, they create in the context of the wide image an unusual beauty.

Carpenter's next film, *Halloween* (the first of five collaborations between Carpenter and cinematographer Dean Cundey), also contains imaginative anamorphic hauntings of the widescreen image via internal framings evoking Academy aperture. The film, about a young babysitter named Laurie Strode (Jamie Lee Curtis) who encounters the murderous rampage of The Shape (Nick Castle) in

FIGURE 4.2 *Halloween* (Compass International Pictures, 1978). Digital frame enlargement.

the small and fictional town of Haddonfield, Illinois, is itself an homage to earlier horror movies, especially those of the 1950s and Hitchcock's *Psycho* (1960). During the evening in which The Shape stalks this sleepy Midwestern town, Howard Hawks's production of *The Thing from Another World* plays on a television set. Carpenter shows us a shot of Laurie and the young boy, Tommy Doyle (Brian Andrews), together on the couch, watching the film. The television plays what Carpenter, on his commentary track on the Blu-ray for *Halloween*, indicates is "my print of the movie [*The Thing from Another World*] on, at the time, one-inch videotape, on my television set, hooked up just for the film" (commentary track, *Halloween* Blu-ray). This shot of Hawks's *The Thing from Another World* is a remarkable example of the expanse of the anamorphic frame finding itself momentarily commanded by the narrower frame of an Academy aperture film, played back on a relatively minuscule television set (with an aspect ratio of approximately 1.33:1; Figure 4.2). Carpenter lets the opening credits of *The Thing from Another World* play out slowly, in accordance with the generally meticulous pace of *Halloween* itself, allowing us to see Hawks's possessive credit (itself forming the inspirational basis for Carpenter's own "writing" of his name above the titles of nearly all his theatrical pictures). The empty space on either side of the television in this shot (likely filmed by Carpenter without the presence of any of the actors) functions very like quotation marks embedding this citation of the Hawks picture within the wider frame of Carpenter's own film. But Carpenter's citation of *The Thing* does not signal any kind of ironic attitude toward the earlier film. Carpenter does not condescend to the earlier Hawks film through this citation but rather earnestly signals his indebtedness to it, so much so that he is willing to focus the viewer's attention in this moment away from the unused areas that are so often elsewhere charged with visual interest in his widescreen frame in *Halloween*, and toward the narrower shape of an earlier film.

This citation of Hawks's *The Thing* is a textbook example of self-reflexive aperture framing, in which the internal frame within the anamorphic image is formed by the reflexive filming of an earlier, Academy ratio movie. This example is also the first significant inclusion of televisions and other approximately 1.33:1 screens within the frames of Carpenter's anamorphic cinema. In later pictures, Carpenter's attitudes toward such screens will be generally skeptical, with televisions and computer monitors functioning as symptoms of social repression, as in the monitors that observe and discipline "criminal" behavior in *Escape from New York* (1981) or in the television sets through which good Americans are taught to dutifully consume corporate product in *They Live*. In *Halloween*, this presentation of Hawks's *The Thing* within the wide frame sets the mood in a way that is simpatico with the kind of atmosphere Carpenter is creating in his film.

It is notable that Carpenter's eventual remake of the Hawks picture, *The Thing*, is somewhat different from *Assault on Precinct 13* and *Halloween* in that it does not contain very many examples of aperture framing within widescreen. The cold exteriors of *The Thing*, shot near the Alaska–British Columbia border, as well as its complex interiors (mostly shot in a studio set), in which a large group of scientists work together in tight quarters, frequently underscore the lateral expanse of the wide frame rather than drawing our attention to aperture framings within the wide image. And what makes Carpenter's monster particularly horrifying in *The Thing* is precisely that it cannot at any point be wholly contained by an anamorphic frame that elsewhere in the film seems the perfect canvas for the depiction of vast swaths of frosty landscape. Carpenter's approach to suspense in *The Thing* is not a matter, as it was for Hawks, of leaving the monster offscreen for long durations of screen time. Carpenter's film, in contradistinction to the Hawks, takes great, gruesome pleasure in showing us its grotesque creature at various points throughout, in the form of the remarkable, and impressively repulsive, figure of the Thing itself, a sculpted mess of goo, animal parts, plastic, and rubber superlatively designed by Rob Bottin, Albert Whitlock, and Stan Winston. On the commentary track to *The Thing* Blu-ray, cinematographer Dean Cundey remarks on the sculptural qualities of the film's monster, suggesting that the various sculptures that constitute different manifestations of the creature's evolution at particular moments in the film are so impressive that they could be housed in a museum of modern art (Cundey commentary track, *The Thing* Blu-ray). The first sustained appearance of the creature in the film implies that it is inherently contingent in its form, its shape and appearance shifting as it mimics the cellular matter of the beings it consumes. The men in the science lab place the creature, which at this moment they believe to be dead, on an examining table. Carpenter's camera moves laterally, leftward, as the group examines it, the frame pointedly unable to contain the entire length of this creature at any single point during the shot. The whole point of the monster is that it can never be fully grasped in any one moment, or stretch of moments, as an ontologically stable being, for it lives to consume the cells of other beings it encounters contingently and ongoingly, and thus the precise nature of its appearance is never final or open to any gaze. The way in which the creature's oozy presence exceeds even the expanse of Carpenter's anamorphic cinema is a palpable expression of the monster's horrifyingly unstable being.

This scene on the operating table in *The Thing* also contains a rare example of a rack focus in Carpenter's cinema; as the camera brushes past the creature's jaunty limbs and oozy red organic material, the gooey carcass, initially the object of attention, is thrown out of focus as the shot moves in on hero Kurt Russell's reaction. But even here, Carpenter's use of the rack focus on Russell is ultimately in the service of guiding his viewer's attention to the larger ambiguities at play in other frames. As Kent Jones writes of the film, "What makes the special effects [in *The Thing*] resonate is the care given to the individual reactions as the Thing undergoes its transformations," expressing "the spontaneous reaction to something hitherto impossible in reality" (30). If Cundy also remarks on the sculptural qualities of the practical special effects used to create the Thing, it is also Carpenter's camera in this film that bequeaths upon the film's creature this (admittedly gruesome) aesthetic quality. Where Hawks, in the earlier version, kept his creature carefully hidden throughout, utilizing our sense of offscreen space, in *The Thing* Carpenter emphasizes a counterpoint between a laterally expansive anamorphic frame and a terrifying creature that never becomes fully manifest to the very type of film camera that would appear ideally suited to capture more and more of the mysterious and contingent reality put in front of it. In Carpenter's film, the complex and terrifying sculptural qualities of the creature, always themselves in a process of becoming and change as it consumes more animals and humans, always ensure that, like a piece of sculpture placed in a museum, we can never quite see in a single moment or from an unchanging position *all* of the object that fascinates and repulses us.

If Carpenter's *The Thing* calls to mind the influence of Howard Hawks on the director's work even as it indicates a markedly different aesthetic approach in relation to the same subject matter, it is worth asking what the reference to Hawks in *Halloween* signifies in relation to that film's widescreen imagery. Again important is the idea of precisely what the wide frame, an ostensibly expansive aspect ratio, can and cannot contain in any given context. What precisely the "horror" in *Halloween* meaningfully constitutes has been a matter of some ideological peregrination and in ways that resonate with this aperture framing created by the presence of Hawks's *The Thing* within *Halloween* itself. Robin Wood has chastised *Halloween*, and Carpenter in general, for not articulating in his film a clear social origin for the evil perpetrated by The Shape in *Halloween*, joining a chorus of critics who find the film reactionary (106–109). Wood, by contrast, in his book on Howard Hawks (written, admittedly, before his shift to a primarily ideological criticism later in the 1970s), seems to approve of *The Thing from Another World* and the complexity he perceives in both the operations of its form and the rendering of its themes and subject matter:

> The climax of the film gains great intensity by this determination to keep us aware of the strength of the opposite position. The Thing, when at last we see it clearly, loses much of its terror. In medium long-shot and from a medium-high angle, it ceases to look huge, and its close likeness to a human being . . . becomes evident. . . . The impossibility of communication becomes almost poignant—it looks as if it would be easy to talk to. It is destroyed: we watch a marvellous, if terrible, being reduced to a small pile of smouldering ashes, on which the camera lingers to allow the spectator a complex reaction. (106)

Wood finds complexity not so much in the way the Hawks picture articulates the origins of its murderous creature—the Thing in Hawks does not allegorically and simplistically "stand for" any particular social evil—but rather in the response of the characters and, by extension, the response of the viewers. Both are enabled to contemplate the reality of the "evil" that has just been banished through the relatively long duration of the shot in which Hawks depicts the smoldering ashes of the Thing. In *Halloween*, Carpenter will also use strategies to obscure The Shape's visual presence to us, although rather than making his monster illegible, Carpenter will frame him as initially marginal, a figure lingering in the corners of the frame or obscurely within it. Sheldon Hall, in his analysis of widescreen style in this film, notes how in

> those shots where Michael [The Shape] does appear as onlooker, he is invariably positioned at the extreme edges of the Panavision frame. These include, for example, his appearance in Laurie's presence, albeit unseen by her, when she peers in through the door of the deserted Myers house, and Michael appears from the shadowy area at screen right and is silhouetted against the light streaming in from outside; his reappearance moments later, as we see a shoulder occupying the same position on the screen, watching as Laurie walks away down the street . . . and Michael's appearance to Tommy's schoolyard tormenter (who *does* see him clearly after running into him, though again we are denied the reverse angle) outside the yard, where he [The Shape] is at screen left, with the boy at screen right. (71)

It becomes almost immediately clear in *Halloween* that despite the outwardly subjective nature of the opening images in the film, we are not meant to identify with the killer. Carpenter's use of the Panaglide, in the first shots ostensibly a representation of Michael Myers's embodied subjectivity (we see the mise-en-scène momentarily through the eyes of a Halloween mask the child Michael wears, the frame itself "masked" to black out everything but two voyeuristic and murderous eye shapes on either side of the frame), ultimately implies the viewer's inability to see "through" the killer's eyes. These shots place us not in harmony with the way Michael sees things but rather at greater points of sympathy with the unfortunate souls on whom he gazes (for a discussion of this, see Hall 70). The fact that he is referred to as "The Shape" in the film's credits also alludes to how the formalistic concerns in *Halloween* supersede characterological ones: this killer is a figural element of cinema before he is a subjectivity. The moniker might also suggest that Carpenter sees The Shape not as a socially produced figure, one whose rampage can be analyzed as the product of a tortured subjectivity the existence of which society is complicit in creating, but rather as, simply, the embodiment of pure evil. The director seems to support this interpretation on his commentary track on the film's DVD/Blu-ray release, which would in turn perhaps support Wood's categorizing of Carpenter as a conservative filmmaker.

And yet The Shape's subjectivity still haunts the film's frames, as an impression that gestures toward a social explanation. If Hawks in *The Thing from Another World* enables us, as Wood argues, to momentarily "linger" on the presence of the vanquished Thing at the end of the film and thereby to contemplate what it is and what has led us to desire its vanquishing, Carpenter has gone beyond even Hawks in this regard in *Halloween*. The entirety of the film, or at least of those shots in which The Shape is marginally present in various ways—lingering at the edges of the frame, partially; emerging from within the depth of the image to spring upon an unsuspecting figure in the foreground of the frame; or slipping offscreen entirely—functions as a means for us to contemplate the alternating presence and absence, and the tortured humanity, of a killer. Unlike Carpenter's Thing in the later film, the *physical* ontology of Michael in *Halloween* is never really in question; we know he is human in some brutal, material sense. It is only the *psychological* ontology, not very much a matter of concern in *The Thing*, that concerns us in regard to The Shape in *Halloween*. It becomes clear as the film goes on that The Shape does not simply "watch" his victims prior to killing them but also watches them after they are dead or as they are in the process of dying, giving us the impression of a murderous subjectivity that, despite all its manifest horror, actually has no comprehension at all of what it is doing. In one particularly haunting shot, The Shape stands in front of the boyfriend of one of the teenage characters, having pinned this doomed young man to a kitchen pantry with a butcher knife. After this gruesome moment, Carpenter frames victim and predator in long shot, from the other side of the kitchen, and at a slight low angle, as The Shape looks up at the young man with a slow, left and right nod of the head, as if in befuddled fascination at the act just committed. Carpenter creates an oblique internal frame in this shot, roughly approximating Academy aperture, isolating killer and killed within vertical lines formed by the edges of the pantry and the shadows. The two figures are also placed at a distance from us, in the middle ground of the frame near the end of the kitchen, as if to implicitly place our own powers of interpretation at a similar distance in this moment of viewing.

I think the effect here is altogether more complicated than simply ascribing to The Shape some kind of unfathomable, eternal evil that eludes sociopolitical meaning. Although Carpenter indeed offers no concrete explanation in this film for precisely how The Shape comes into being, his film paints a picture of a larger social failure to understand what is, undeniably, a particularly horrible and, as the film goes on, horrifically legible type of subjectivity. When The Shape does appear onscreen more or less fully—"fully" here relative to shots in which only part of his body is viewable in the frame—Carpenter's method of internally framing this figure does not just create a great deal of suspense and anxiety (it is precisely the fact that The Shape *does not* command the entirety of the frame that makes him scary to us, for we fear he will soon do so). It also serves to bracket The Shape from the surrounding social mise-en-scène in *Halloween*, from all of its institutions that have failed to discipline, control, and understand him.

The Shape, after all, is not just The Shape, a figure at play in Carpenter's frames. He is also Michael, possessed of a terrible humanity, and as such is capable not only of subjectivity but also of playing with the way he appears to others. Most of this appearance takes the form of an expressionless white mask, an image perhaps of a kind of terribly misdiagnosed or untreated autism, that he wears throughout the film. (We see Michael's face only once and very fleetingly, near the end of *Halloween*, prior to his being shot just before he puts his mask back on.) But Michael is capable of shifting appearances in relation to what he sees. Apparently befuddled by the aforementioned death he has caused in the family kitchen, when Michael next appears, it is upstairs to the view of the girlfriend (P. J. Sloane) of the murdered boy downstairs. Here, in one of the film's memorable and hauntingly playful images, Michael appears as if he were the ghost of the young man he has just murdered, dressed in a bedsheet and wearing the boy's glasses. The young girl, in bed, thinks it is her boyfriend, goofing around. Carpenter's shot of Michael once again relies on an internal frame that again evokes Academy aperture, presenting this ghost variation of Michael, still for a moment as he stands in front of the girl, from within the vertical lines formed by the bedroom doorway. This internal framing of Michael onscreen echoes the internal framing in the earlier shot, with the stronger vertical lines of the bedroom wall now taking the place of the shadows and kitchen pantry. But Carpenter's playful variation of internal framing is also itself a depiction of Michael's hauntingly, if obscurely, self-conscious variation of his own postmurder befuddlement. In the earlier shot in the kitchen, the internal frame was a representation of Michael, a type of oblique aperture framing in which the killer was taken as the object of the frame's investigation. In the subsequent shot of Michael-as-ghost in the bedroom, Michael is now suddenly, terrifyingly, and seemingly subjectively aware of this power of framing, now presentationally appearing, as if this killer were suddenly invested with the powers of theatrical self-creation, as a ghost character he has incarnated in order to try to "understand" the young man he has murdered. This presentation of the self-as-ghost, the moment suggests, is Michael's way of understanding, indeed arcanely inhabiting, the very figure of the dead body for which he is responsible in the earlier scene but which he also apparently fails to understand as either a subjective or ontological entity.

And so Michael becomes legible as a kind of distant subjectivity here, through this rhyme of internal frames, in which he seems to take on a kind of dim and terrible self-awareness. The internal frame, as composed by Carpenter, implies also the ongoing way in which this killer, isolated by society, has also been produced by it. Carpenter has also retained and reminded us of the power, from within the context of his deployment of Panavision, of aperture framing, which continues to haunt the wide frames of *Halloween* in compositional ways and which his citation of Howard Hawks cues us to look for.

Televisual Parasites in Panavision: *Prince of Darkness, They Live*, and *Memoirs of an Invisible Man*

As moments from *Assault on Precinct 13*, *The Thing*, and *Halloween* suggest, Carpenter's Panavision cinema engages playfully with earlier, narrower aspect ratios, image-making that haunts the director's frame in forms of television sets, aperture framing, and internal frames within the anamorphic image. In later Carpenter films, narrower aspect ratios, by contrast, generate acute anxiety about cinema's own relation to competing, and reductive, formats of viewing and image-making. In a trio of films stretching from 1987 to 1992—*They Live*, *Prince of Darkness*, and *Memoirs of an Invisible Man*—Carpenter's citation of home video, and narrower aspect ratios in general, becomes the very threatening forces besieging the protagonists, paralleling the existential threat to cinema formed by home video. In his study of VHS and its impact on the film industry, Frederick Wasser points out that video triggered a change in film style during the 1980s and early 1990s, the period in which these films were made. Building on Jerry Mander's suggestion that television imagery deploys "technical events" in a bid to appear more cinematic—"a cut, a zoom, a superimposition, a voice-over, the appearance of words on the [TV] screen . . . each alteration of what would be natural imagery" (Mander qtd. in Wasser 197)—Wasser cites the tendency of many films of this period to add superfluous techniques to the cinematic image to ensure that it "plays better," and retains a certain kind of pseudo-cinematic aura, when transferred to the relatively impoverished, and necessarily narrower, playback format of the VHS tape. "As the shots are simplified so that they play on the small screen," Wasser goes on to write, "the filmmaker starts adding technical events to compensate for the simplification" (197).

Carpenter's films, notably, eschew such "technical events." The director remains committed to anamorphic widescreen Panavision during this era of pan-and-scan transfers of films on VHS. Nevertheless, in the trio of films under discussion in this section, Carpenter explores the fate of the Panavision canvas in relation to the intrusion of videotape and other bastardized forms of home viewing. In *Prince of Darkness* and *They Live*, video and television are the vehicles by which parasites and invaders announce their presence to the protagonists. These low-resolution video images are in turn thrown into relief against Gary B. Kibbe's astonishing cinematography on these two films. In *Prince of Darkness* and *They Live*, video and television function as parasites taking over the host of Carpenter's Panavision frame, paralleling the ways the characters find themselves under siege by supernatural or otherworldly forces in both films. In *Memoirs of an Invisible Man*, the memoirs of the title are recorded by the invisible man on videotape, the source of the film's ensuing voice-over (a rarely used technique in Carpenter's cinema) narrating the flashbacks. In *Memoirs*, Carpenter's Panavision frame shows us more than the impoverished videotape on which the invisible man records his memoirs could possibly display, the film's "conquest" of the narrower aspect ratio paralleling the invisible man's discovery of his own ability to move across the anamorphic frame with agency and heroism. To some extent, *Memoirs* in this way assuages the anxiety over the intrusion of home video into the frame of Carpenter's anamorphic cinema. Nevertheless, the film's own pan-and-scan release on videotape, the fate of all of Carpenter's widescreen films in the 1980s and 1990s, ironically reverses this victory.

Prince of Darkness

Prince of Darkness, the first of Carpenter's seven collaborations with cinematographer Gary B. Kibbe (the two will work together again on *They Live*, *In the Mouth of Madness*, *Village of the Damned*, *Escape from L.A.*, *Vampires*, and *Ghosts of Mars*), combines religion and science in its story of a priest (Donald Pleasance) working with a team of PhD researchers to combat the arrival of a demonic spirit in the underground catacombs of a California church. This spirit—which Carpenter himself refers to as an "anti-God" (Boulenger 201), evoking the relationship between matter and antimatter in quantum physics—manifests as green, gelatinous ooze, percolating in a cylindrical container in an underground crypt monitored by clergy for centuries. As the scientists gather in the church, they assemble computer monitors on which they will work to translate the cryptic messages of ancient books that foretell this spirit's arrival. The square shape of the computers is another example of a visual form thrown into relief against Carpenter's wide frame. But the 1.33:1-display technology most important to *Prince of Darkness* is evoked in the videotaped images of the church exterior, through which we spy a demonic figure lingering in the doorway. These unsettling, low-resolution video images become the film's haunting motif. Each time they appear, they function as representations of the spooky "dreams" of various characters, dreams apparently connected to the demonic spirit in the green ooze. As the narrative goes on, it is explained that these "dreams" are telepathically sent recordings from a future humanity attempting to warn the present-day characters in *Prince of Darkness* of the impending evil the demonic spirit is about to unleash. Regardless of how the story rationalizes the origin of these video images, what remains most memorable about them is not their function in the narrative but the way in which they come to haunt, to literally "take over" for a stretch of time, the entirety of Carpenter's widescreen frame. The way in which the demonic spirit comes to parasitically feed upon the characters in the film, a process by which it is ingested into their host bodies, is paralleled on the visual level by how the videotaped images of this spirit come to take over the entirety of Carpenter's anamorphic frame.

These videotaped images are notable for several reasons. First, in their garish, rather overexposed, low-resolution representation of the church, they stand in counterpoint to cinematographer Kibbe's delicate, painterly yellow, and beautifully lit imagery in the interior basement of the church where most of *Prince of Darkness*'s unsettling events occur. As if to suggest a shared visual sensibility between themselves and their characters, Carpenter and Kibbe have cleverly connected their own cinematographic craft to the methods of the researchers. When the PhD students and their professor install themselves in the church and its secret, underground lair to begin their study, they arrange around the catacombs, surrounding the many candles already lit in the church, several large lamps kin to the illumination Carpenter and Kibbe use to burnish their own frames. The effect is that the narrative space of *Prince of Darkness*, a place of careful scientific research and philosophical, theological speculation, comes to resemble, with lights strewn laterally across the frame, something like a movie set itself. This luminosity makes the occasional appearance of the rough, lower-resolution videotaped imagery—intruding as it does into the dreams of the characters and across the entirety of the film's own Panavision expanse—even more jarring.

Second, Carpenter does not attempt to use this videotaped imagery like a "technical event" intended to generate some aura of "cinema" in the film's eventual transfer to home video; the appearances of these videotaped dreams do not augment the Panavision image in an artificial or superfluous way. The videotaped imagery that consumes Carpenter's anamorphic frame at certain points is neither particularly attractive nor especially compelling *as imagery*; it is haunting only in the context of its nerve-jangling intrusion into the celluloid cinematography of *Prince of Darkness*. More important, the imagery is presented as a *reduction* of familiar cinematic technique within the wide frame, rather than a sign offered by Carpenter that videotape could possibly add to what was already possible in cinematic widescreen. On the commentary track to *Prince of Darkness*, Carpenter notes that he and his crew recorded the videotaped imagery appearing in *Prince of Darkness* and then recorded that imagery off of a television that played on set (*Prince of Darkness* Blu-ray). The composition of this imagery, in at least this sense, is admittedly quite cinematic, or at least informed by an idea of the cinema. Each time it appears, we see a variation of a similar thing: the camera, hiding behind a white fence, floats, handheld, to the entrance of the church, where it spies the looming, dark figure of a haunting spirit. In this sense these videotaped images are produced using the same cinematic ideas Carpenter employs elsewhere in his anamorphic filmmaking (handheld camera movement, invoking the presence of a subjectivity; the presence of a figure in the background of the frame). But via the videotape these techniques now take on a low-resolution visual form. The presence of the church's name in these videotaped images—"St. Godard"—also invokes the name of a major auteur, Jean-Luc Godard (see also Cumbow 164), and suggests the spirit of cinema living on, even as it is "contained," in these images in the impoverished form of low-resolution videotape. Importantly, the frame formed by the screen of the television itself on which Carpenter and his crew play and record the videotaped footage does not appear in any of these images of the videotaped footage in the dream sequences; unlike the presence of the televisions in *Halloween* (or, as we will see, in *They Live*), in which a TV and its broadcast are presented as part of the social world of the film's characters, the videotaped imagery in *Prince of Darkness* has been extricated from its source and now floats through character psyches in a haunting and disembodied way, even as it takes over the entirety of the wide screen and even as it warns the characters of the impending presence of evil.

Third, and perhaps most intriguingly, Carpenter uses videotape's low-resolution engulfment of the anamorphic frame as a way of invoking the idea of futurity (in a film that contains elements of science fiction), and by extension also the impending mortality of celluloid-based Panavision as both a material entity and an available aesthetic choice within the film industry. One theory shared by the characters in the film is that these video images are kinds of "telepathic signals" sent from future human beings to warn our present-day characters that the evil spirit housed in the catacombs of the church has succeeded in its intent to take over human beings as hosts. On the narrative level, these dreamy video "messages" help the protagonists: they warn them of a future in which the evil spirits have taken over humanity. However, the appearance of these videotaped images is somewhat separate from the theoretical explanation of them offered by the characters. When they first appear, we do not know they are meant to signal some kind of "future message" from benevolent figures; even when the images are repeated later in the film and upon repeat viewings of *Prince of Darkness*, the appearance of the videotaped image of the looming demon inside the church takes on the ominous, anxious quality of a threat, as if these images were sent by the demon itself. This uncertainty and anxiety over the mode of address of the VHS-based televisual image in *Prince of Darkness* emerges through that image's tension as it courses through and is ultimately contained by the anamorphic frame. This feeling is most palpably felt as a matter of one type of imagery invading another, as this jarring and jagged intrusion of a VHS aesthetic into the widescreen of *Prince of Darkness* is perhaps also the unconscious expression of an auteur's own anxiety in relation to a home video format that frequently sliced and diced his screen compositions on their home video editions during the era in which this film was made. If the overall haunting quality of *Prince of Darkness* comes, then, not only from its clever play with cinematic space but also from the palpable tension the film presents between Panavision celluloid and the parasitic invasion of videotape into the widescreen frame, this is an aspect that is perhaps more legible today than it was for many of Carpenter's viewers in the late 1980s, who likely encountered *Prince of Darkness* for the first time on pan-and-scan home video or pay cable.

They Live

They Live exudes a similar anxiety over the presence of the narrower frame within widescreen compositions, an anxiety Carpenter redirects into stylistic composition and political commentary. The film's premise is that alien beings have colonized Earth, delivering brainwashing, subtextual messages via television signals intended to keep humans docile and happy as they gobble up television programs and consume the advertised corporate products. The resistance to this exploitative domination occurs, in the narrative, via television, as political organizers in the film—via interruptions of the "official" television signal with their own transmissions—counter and resist the ideological message of the aliens. In interviews and commentaries, Carpenter is fond of describing *They Live* as his only explicitly political film, one meant to resist the unregulated capitalism and repressive ideology of Reaganomics in the 1980s (see Boulenger 209). Where, then, is Carpenter's cinematic practice itself situated in terms of the "resistance" to a consumerist lifestyle, ostensibly fought in the narrative via television? Tellingly, the resistance fighters in *They Live* seem to know that interrupting the television signal in a bid to "wake up" the sleeping denizens of the capitalistic colony will not be enough to defeat the aliens. They also come up with optical inventions that affix themselves to the human eye and remind us of other cinematic and visual technologies—first, glasses (evoking the use of 3D glasses in viewing some forms of spectacular cinema) and, then, later in the film, contact lenses, enabling the heroes to see the aliens among the human population. These glasses also enable the film's viewer to "decode" the mise-en-scène to reveal the existential toxicity of consumer capitalism. Carpenter's implied belief in *They Live* is that cinema, and not narrower or more atomized forms of visual communication, will teach viewers how to resist totalizing ideological meaning, precisely through a free perceptual and creative play with wide images.

Only through following the adventure of Carpenter's protagonist (played by Roddy Piper, and only retrospectively named "Nada" on the end credits—his name is not mentioned in any dialogue in the film, only underscoring his identity as "nothing") will we discover we can only see the presence of the alien beings once we learn how to see the wide frame through those glasses. And rather than engulf the entirety of the Panavision frame with the televisual image as he does in *Prince of Darkness*, Carpenter situates television, and its various brain-altering signals, as a concrete part of *They Live*'s mise-en-scène. Much of the film's early

FIGURE 4.3 *They Live* (Universal Pictures, 1988). Digital frame enlargement.

stretch of action takes place in a shantytown called Justiceville, and the first television sets we see in the film, broadcasting both the aliens' consumerist message and the resistance offered by the anti-alien political organization, are those around which some of the residents of this makeshift community gather to watch at night. Carpenter presents these televisions in his anamorphic frame much as he presents the television on which Laurie and Tommy watch the Hawks picture in *Halloween*, although now the presence of a narrower frame in the context of Panavision registers more as ideological warning than as auteur homage (Figure 4.3, another example in Carpenter of aperture framing).

As Jonathan Lethem notes in his monograph on the film, the opening sequence of *They Live*, far from conforming to the convention of creating images legible on both cinema screens and home video, constitutes a kind of cinematic tutorial showing a viewer weaned on pan-and-scan VHS how to engage again with creatively orchestrated widescreen imagery. Lethem notes salient moments of apparently intentional "optical confusion" in the first thirty seconds of *They Live*, involving the appearance of the film's title card, stylized as street graffiti, before the leftward movement of the camera confirms the spray-painted title to be part of the narrative's urban, Los Angeles reality. Even the leftward movement of the camera is complicated by "the rightward drift of the train cars" that Lethem observes, the opening shot testifying not only to the surplus of pro-filmic reality existing in front of the apparently expansive frame of a Panavision lens but also to Carpenter's deft ability to use this space to both artistic and pedagogical advantage, tutoring the viewer in the perception and interpretation of widescreen images (Figure 4.4). If one sees *They Live* in a cinema or on a disc that retains the original aspect ratio of the Panavision image, one receives what Lethem characterizes as "a warning—matters of competence in 'reading' images will be at stake here" (10). But most viewers—during the late 1980s and 1990s, at least—likely saw *They Live* on VHS, rendering moments of "optical

FIGURE 4.4 *They Live* (Universal Pictures, 1988). Digital frame enlargement.

engagement" like this, which do not quite achieve the status of the "technical events" used so often during the era to charm viewers of VHS into perceiving televisual images a "cinematic," unintentionally ironic.

The material history of *They Live* itself as an artifact both cinematic (on celluloid and on letterboxed Blu-rays, DVDs, and laserdiscs) and televisual (as a pan-and-scan tape) becomes, then, rather unexpectedly a part of the theme of the film itself. I saw the film for the first time on VHS in the early 1990s, and when I finally saw a letterboxed version on DVD later in the decade, the effect of the revelation that there was *more* image to see and engage with, and indeed that a certain kind of "scanning" and "panning" of the narrower image had in fact substituted for my engagement with its own televisual intrusions, was somewhat similar to the Roddy Piper character's revelation when he puts on the sunglasses. (Later repertory screenings of the film on 35mm only threw this revelatory expanse into more vibrant relief.) I could, effectively, with these later screenings, be enabled to "read' the film for the first time, Carpenter's composition of screen space and the viewer's attendant attention suddenly becoming an important part of the experience. Many of the most striking images in *They Live*, above and beyond the startling revelation of the alien control of reality upon Nada's discovery of the vision-altering sunglasses, are the relatively more subtle shots that, when read in their original 2.35:1 composition, reveal details repressed in the film's pan-and-scan version.

It is worth looking at some of the frames from the film that suffer the most in the pan-and-scan process when thinking about the subject matter of *They Live* in relation to its use of the anamorphic frame. Many such images involve a composition in which important information is present, during a stretch of screen duration, on both the far left and far right sides of the frame, the type of composition that on a pan-and-scan VHS tape would either be artificially "re-edited" (with the single anamorphic shot "cut" into an artificial shot–reaction shot in the videotape transfer) or "scanned" (an electronic "pan" moving across the surface of the narrowed frame), introducing a technical event into the image that effectively substitutes for the viewer's own creative gaze. Such a composition occurs almost straightaway in the film's titles sequence, in a shot of Nada walking alongside an immobile, rusting train, which extends diagonally across nearly two-thirds of the frame, as automobile traffic in the street sweeps by on the left. A shot that in a tighter, panned-and-scanned composition might simply be read as "a drifter walking" becomes, in its original anamorphic composition, more complex. Nada, in the wide image, is himself situated in-between, and is visually registered as relatively minuscule in relation to, the busy labor and consumption of those who can afford to buy and maintain cars (on the left side of the frame) and decaying locomotive industries, signaled by the abandoned train (on the right).

Another such complex anamorphic image, also featured against the opening credits, is one of the many involving the several homeless people who serve as "extras" in the surrounding areas of Carpenter's frame, not only giving his imagery of Nada's experience as a drifting, working-class man authenticity but also inscribing the citizens on the streets of Los Angeles as a potential constituent of the film's political message (on the commentary track to the Blu-ray, it is revealed that these people were paid as extras for the day when Carpenter shot in the street locales in which they were present). The presence of many such extras is elided and discarded as, presumably, narratively unnecessary "information" when the film is cropped to 1.33:1. There are many such images throughout the first stretch of the film, in which Nada or another of the main characters in Justiceville is seen in long shot, as figures linger in the background of the frame, citizens of Los Angeles, some of them presumably without homes, who have joined Carpenter on the shoot.

The presumed authority, the false authorship, of those who pan and scan widescreen films into VHS transfers to determine what is at any moment "narratively necessary" subject matter is in this way reflected upon in the ideological operations of the film itself, the narrative of which involves just such resistance against images infused with a consumerist, rather than expansively aesthetic, sensibility. The first stretch of *They Live* is peppered with images of the very kinds of televisions on which viewers during the late 1980s and early 1990s would have viewed the pan-and-scan tape of *They Live*, the film's inclusion of these 1.33:1 displays functioning as a kind of acknowledgment of the ghoulish afterlife of anamorphic imagery during this period of infrequent letterboxing on home video. Of course, on the VHS transfers of *They Live*, "visual priority" in the panning and scanning is given to the presence of the film's main actors (Piper, Keith David, and Meg Foster, unlike the extras in the backgrounds of Carpenter's frames not seen by the pan-and-scanners as visual excess that can be "cropped" quite so easily), and so it could be argued that this repression of ideologically oriented meaning (such as that within the two shots I have pointed to earlier) is

FIGURE 4.5 *They Live* (Universal Pictures, 1988). Digital frame enlargement.

an effect of the film itself and its conventional reliance on a certain hierarchy of star performance, with pan-and-scan only underscoring an ideological facet of the existing motion picture. Other compositions in the film, particularly those not featuring any of the film's main stars, suggest that this is not so, and that Carpenter's use of the anamorphic frame is tied into the ideological thrust of his film. The most striking of these are of course the images of the ontological reality that Nada uncovers when he, and by extension we, put on the glasses. One such frame presents a laterally arranged vision of various subliminal, consumerist messages encoded in the various advertisements dotting the streets of Los Angeles (Figure 4.5). These messages—"Obey," "Consume," "No Thought," "Marry and Reproduce," and similar others—are seen in other such shots in the film, but only in this one image are nearly all of the hidden messages arranged together in this way, the shot functioning as a kind of neat summation of the ideological substance implanted in human society by the aliens. In the film's narrowed pan-and-scan version, some of these messages are elided entirely in this shot, suggesting the way in which cropping the anamorphic image can repress a film's ideological complexity. Significantly, Nada's mission ultimately involves destroying the transmitter emitting the aliens' signal from a cable television news network—the same signal that implants these messages and that is encoded in the 1.33:1 images represented throughout *They Live* in the various television monitors arranged in the mise-en-scène.

Memoirs of an Invisible Man

If *They Live* situates televisions—or, more generally, narrower aspect ratios—as sites of existential and ideological repression, Carpenter playfully complicates the relationship between frame shape and subject matter in his next film. *Memoirs of an Invisible Man* is the director's only collaboration with veteran Hollywood cinematographer William A. Fraker, whose celebrated work also includes

Rosemary's Baby, *Bullitt*, *Exorcist II: The Heretic*, *Looking for Mr. Goodbar*, *Old Boyfriends*, *WarGames*, *The Freshman*, and *Rules of Engagement*, among many other films. Inverting slightly the cinema-videotape dialectic of *Prince of Darkness* and *They Live*, in *Memoirs of an Invisible Man* the anamorphic image becomes a stage for deception and lies, as the diegetic government works to cover up the scientific accident that results in our hero's invisibility. It is the hero's freshly invisible ontological status that one rogue CIA agent, Jenkins (Sam Neill), wants to appropriate in order to turn Nick Holloway (Chevy Chase) into a kind of superspy (one whose services Jenkins plans to sell on the open market). By contrast to the deceit and espionage coursing through the wide image of the film, the ostensibly impoverished medium of the videotape, on which are recorded the memoirs of the title, serves as a site of ostensible truth in a narrow frame. Although initially glimpsed in the recording the invisible man is making on a video monitor in the film's opening sequence, this testimony ends up floating throughout the ensuing movie in the form of Chase's voice-over narration (much of the main body of the film is related in flashback). But even that straightforward notion—the idea that the anamorphic frame is the container for deceptive narrative complications that our invisible man will put to rights in his earnestly videotaped, narrow-ratio memoir—is complicated by the larger operations of the film itself. In Carpenter's hands, *Memoirs* wants to demonstrate that the wide cinematic frame, as late as 1992, is still potentially a canvas for visions and wonders that simply cannot be replicated without diminishment on the smaller, narrower screens of a home video recording.

In a shot in the film's first sequence after the opening title credits, the camera tracks toward an apparently empty chair at a desk in an empty electronics store after closing hours, with a video camera set up on the opposite end and a row of monitors behind the chair replicating the image of the chair being recorded. This shot is a neat demonstration of how much of the scene, in terms strictly of narrative legibility, remains intact even when viewed on a narrower video monitor: we can see the "entire" space on those video monitors, a demonstration simultaneously of what is lost when the wide image is narrowed down to 1.33:1 but also of what remains substantially legible even after such diminishment. Of course, this chair is not really bereft of presence: the invisible man is soon revealed to be sitting in it, as he tells us through the undiminished power of his voice, and proceeds to provide as proof of his invisibility a demonstration of what to our eyes looks like, first, a pencil and an eraser floating in air, and then, a piece of bubblegum apparently unwrapping and then chewing itself. This demonstration only occurs after the camera has tracked in to a "close-up" of Nick, however, putting the aforementioned television monitors that duplicate his "image" in the mise-en-scène and video camera offscreen as if the film were suggesting that a spectacular feat of visual effects were only possible in cinema. Whatever agency Nick has lost in becoming invisible in his world (and *Memoirs* will be particularly notable for how it stresses the unattractive rather than advantageous or alluring aspects of this "power"), for us his invisibility renders him a spectacular being, a creature possible only in movies. But as our earlier glimpse of the narrower video monitors has suggested and despite the cinematically spectacular nature of this special effect, invisible Nick here is presently taking up only a small slice of the anamorphic frame. A row of videocassette players remains visible behind—indeed, *through*—him, reminding us that while the diminishment of the effects we are seeing will be unavoidable when the film is panned and scanned to VHS, the action is nevertheless quite centered in the frame and as such would not be made entirely illegible on that format.

But narrative legibility (our ability to read the image, even on a videotape, for its to-the-point "narrative information") is not quite the same as achieving narrative agency—the character's, in this case Nick's, existential control over what can and might happen to him. Nevertheless, there is something of a parallel on the level of the character's experience for our creative negotiation of what a film at any given moment might mean to us. To take control of his situation, Nick will have to move out of the small quadrant of safely squarish screen space within the wide image (such as that which describes his position at the desk above, which is in fact his position throughout most of the film as he relates events in flashback to us), a space compositionally designed for the monitor on which he records his memoirs, his testimony, which ostensibly can still be "heard" even when *Memoirs* is viewed in pan-and-scan. In relation to those memoirs, recorded during a stretch of time when the invisible man safely locks himself away in the electronics store—and "safely" within the narrow confines of the 1.33:1 monitor on which is shown the video image of his testimony being presently recorded—the anamorphic space surrounding him takes on an uncertain, dangerous tenor. It is a space that, once his recording of the memoir is finished at the beginning of the film's third act, he will eventually have to venture out into again in order to regain some measure of existential control over his life.

Nick's voice-over floats throughout the film as a kind of reminder that, even as we see Chase enact events in Nick's life set in the ostensible past, the character himself remains seated, relatively safely, in that chair carefully placed within the slice of narrowed reality available to the 1.33:1 monitor as he relates his widescreen adventures to his listener. The voice-over is a kind of floating memory of that narrow margin inhabited by invisible Nick seated in the chair in the opening sequence. Michel Chion, in writing of the voice-over that recounts narrative events, remarks that this technique "designates any acousmatic or bodiless voices in a film that tell stories, provide commentary, or evoke the past. *Bodiless* can mean placed outside a body temporarily, detached from a body that is no longer seen, and set into orbit in the peripheral acousmatic field. These voices . . . quickly find themselves submerged by the visible and audible past they have called up—that is, in flashback" (49). Even the flashbacks themselves, although "in the past" and not posing any direct threat to Nick, demonstrate repeatedly, for us, the perils of his invisibility and frequently make us forget that we are watching something that, in the film's time frame, has already happened. As action sequences, they temporarily submerge our knowledge that Nick, throughout most of this movie, is in fact sitting alone in an electronics store, recounting all these events, which he has already survived, into a low-resolution video camera on a 1.33:1 image, even as we see them unfolding in and across the expanse of the anamorphic image.

The film repeatedly stages events, both in the flashbacks and in the sequences in the third act (after he finishes recording his memoirs in the electronics store and is once again under threat of discovery by the film's villains), in which Nick, safely positioned for a time in a narrowed slice of the wide image, must daringly act and move out of narrowed, safe confines, careful always not to breathe or move about too loudly. The antagonists in the film at times wear heat-seeking goggles, enabling them to detect Nick's presence (a kind of inversion of the glasses in *They Live*, given that the glasses are now being worn by the authorities in power); Carpenter shows us their viewfinders upon which Nick is reduced to the dabs of red that indicate, through the ineluctable confession of his body heat, his position in space. But the villains do not always have those goggles at hand, and during such moments they must grapple with the way Nick is still able to use his human form to navigate anamorphic space.

One neat little demonstration of the potential perils awaiting Nick when he dares to move out of the narrow confines of one slice of the wide frame occurs when he hides in Jenkins's office, with the intent to find information that might incriminate the rogue CIA agent. As Jenkins works at his desk, unaware of the presence of the invisible man, Chase—presently visible, in a technique the film will occasionally use to convey the presence of Holloway to the viewer, at screen right, even though he is invisible to himself and others in his world—huddles, knees bent against his chest, leaning against a file cabinet in the corner of the office. Most of screen left is empty office space; Jenkins is offscreen. At one moment, Nick stretches a leg out to relieve a cramp. An ankle bone cracks. Carpenter's camera, at this, begins to track in toward Nick: the very screen space he has sought to hide within is now enclosing upon him, and he will have to learn how to move within it and across it if he is not to be vanquished by it. Jenkins, having heard the ankle crack, now knows Nick is here and pushes a button hidden under his desk, locking the door to his office. Our hero must now inhabit the entire stretch of the anamorphic frame with improvised creativity, navigating the very sort of life-threatening situation he has hitherto worked to avoid. Chase begins moving around the room in a cat-and-mouse game, using both his voice and his invisibility to variously clue Jenkins into where he is standing before then moving to another slice of screen space in order to ambush him. In the sequence, he eventually ends up circling across the space around Jenkins's desk, successfully moving across the lateral stretch of screen space in order to nab Jenkins's gun, hover it invisibly next to his antagonist's head, and make his way out of the office.

FIGURE 4.6 *Memoirs of an Invisible Man* (Warner Bros., 1992). Digital frame enlargement.

Nick must learn to act and move across the widescreen frame, which of course defines the work of any action hero but becomes an even more intense form of labor when one cannot be seen. Above and beyond any action undertaken by the characters in Carpenter's mise-en-scène, however, it is the technical achievement of the film itself, which is both an astonishingly mounted homage to the various classical Hollywood renditions of the archetypal Invisible Man story (at one point, when revealing his identity as such to Daryl Hannah's character, Nick undoes facial bandages with a gesture and in a costume that reminds us of the "presence" of Claude Rains in the 1933 version) but also a technical feat in and of itself. Carpenter and his team devised a camera setup, guided by a motion-control unit, that could exactly replicate the movement of the camera in the filming of shots in which the viewer is intended to both see Nick and not see him, as in one shot in which Nick is visible to us in the frame even as he is invisible in the reflection of a mirror behind him (Figure 4.6). As the film's cinematographer, William A. Fraker, explains, the lighting of the shots in which Chevy Chase is variously visible and not visible required four times as much lighting as when he is filmed in the frame, owing to the differences between the light setting needed to light an actor and the settings required by Industrial Light and Magic to implement its special effects (Gentry 46). This requirement was in turn complicated by Carpenter's preference for wide-angle lenses paired with anamorphic cameras, a combination that demands more intense arrangements of light even in conventional circumstances. After the resulting "mirrored" images are composited into one final shot for the finished film, one can both see and not see Chase seamlessly, a kind of correlative to the way in which the villains can both sense and not sense him during scenes in which they know he is present.

If Nick finds his agency in learning how to move across the anamorphic frame, the result is nevertheless one of the most conservative final sequences of any Carpenter film, largely owing to the studio's suppression of the originally shot ending, which involved the birth of a potentially monstrous invisible baby upon the union of the Chase and Hannah characters (Carpenter did not have final cut over *Memoirs*). Warner Bros., worried that viewers would be disturbed by such a conclusion, forced Carpenter to cut the sequence, and what remains is an anodyne romantic clinch between the two lovers after Nick's various complications are worked out (he nevertheless remains invisible, a mild subversion of the ostensibly happy ending). Despite this ending, the film, as a widescreen work of art, works best when Nick is kept still in the frame and Carpenter is able to use the wide image to poetic and painterly ends. The most memorable imagery in the film takes on a surreal quality, such as the depiction of the office building in the beginning of the film that is the site of the scientific accident that renders Nick invisible. The building itself is rendered partially invisible by the accident, and in Carpenter's imagery we can see slices of the building interior as if "floating" in space against a backdrop of blackness. The effect of such frames is akin to a surrealist painting by Salvador Dalí, wherein time becomes tangible in objects, floating and bending in ways that violate our preexisting sense of the material world. "I do like Dali and Magritte a lot," Carpenter says. "I like surrealists because I find them funny, amusing, and anarchist. And the conception of that invisible building with this kind of melted-watch look was really based on that. It reminded me of Hitchcock's *Spellbound*" (Boulenger 222).

Other such surreal images in *Memoirs*, which apparently exceed in their strangeness the ability of the invisible man to record them as spoken facts in his memoir recorded on the video camera, include a romantic evening scene in which Chase and Hannah, on the run from Jenkins, have an interlude on a drizzly, wet evening. The contours of Nick's body are rendered partially visible by a magical coat of rain. Another surreal sequence involves Hannah rendering Chase's flesh manifestly visible through the application of makeup, this time in a strikingly heightened, even feminine form. In this image, a visible slice of Chase's painted head floats in the middle of the anamorphic frame, the strangeness of this suspended facial figure in the middle part of the image throwing into relief the space surrounding it (Figure 4.7). Such playful imagery literally defigures the classical Hollywood hero and uses him as the source for visual rather than strictly narrative play within the wide image.

Reflections on Authorship: The Presence of the Author in *The Fog* and *In the Mouth of Madness*

Prince of Darkness, *They Live*, and *Memoirs of an Invisible Man*, each in their way, find Carpenter's anamorphic cinema haunted by the parasitic and altogether unsettling presence of narrower aspect ratios that form an existential threat to the wide image, the foundational canvas for Carpenter's signature as an artist in cinema. These later films are distinct from *Assault on Precinct 13*, *Halloween*, and *The Thing* in their refusal to engage in a dialectic between wide cinema and

FIGURE 4.7 *Memoirs of an Invisible Man* (Warner Bros., 1992). Digital frame enlargement.

narrower aspect ratios, preferring instead to see the latter as a kind of threat that must be conquered or contained. I want to end this chapter's study of Carpenter, and this book, with a focus on two films that set aside this acute anxiety over other forms of media in favor of a more direct contemplation of film authorship and artistry in the wide image. These two films do not trouble themselves with narrower aspect ratios but rather plunge headfirst into the dizzying potential of the wide frame. *The Fog* and *In the Mouth of Madness*—for me, Carpenter's two greatest films—tell chillingly good and horrific stories, reminding us again of Kent Jones's observation that Carpenter is a great genre filmmaker before he is anything else. But, taken together, they also constitute Carpenter's magisterial reflection on authorship in widescreen horror.

The Fog

The Fog (Carpenter's second collaboration with *Halloween* cinematographer Dean Cundey) begins with one of the more memorable opening shots in Carpenter's oeuvre, an image that intimately links gestures in the frame to those of the film author. A storyteller, Mr. Machen (John Houseman), sits by a crackling orange fire and in the surrounding darkness relates a spooky tale of drowned sailors to a group of enraptured children. Our introduction to this storyteller is not through his words, however, but through the object of his pocket watch in the foreground of the frame in close-up as the film begins. With Carpenter's atmospheric, minimalistic soundtrack scoring the shot, the camera tracks slowly leftward toward the listening children as the pocket watch, still dangling from above the upper right side of the frame, follows along with it (Figure 4.8). Here, the gesture of the storyteller and the rapt attention of his audience are figurally rendered across the length of the Panavision frame, each to its separate side (storyteller on the right, listeners on the left), as if suggestive of the way cinema requires both a director and a viewer to make productive sense and experience out of the

FIGURE 4.8 *The Fog* (AVCO Embassy Pictures, 1980). Digital frame enlargement.

entire breadth of the image. When the camera stops on the children, the pocket watch dangles for a moment, before a hand reaches from below and clamps it shut. The effect joins the gesture of the filmmakers (the movement of the camera) with the gesture of the storyteller—the presumed Mr. Machen, holding the watch and controlling the narrative, at least for now. Like Carpenter himself, Machen is, for the first stretch of this shot when he holds the watch in front of our eyes, invisible, the offscreen narrative agent shaping and authoring the frame that creates our sense of story. After Machen tells of the drowned sailors who will come to seek revenge against the descendants of those who one hundred years ago led them to their deaths, Carpenter's camera moves upward, past Machen and toward the foggy horizon from where the pirates, seeking to claim their justice, will soon emerge.

Carpenter himself will appear in an early scene as the janitor in a church, engaging in one brief exchange with a priest, Father Malone (Hal Holbrook), a relatively rare appearance in front of the camera for a director not very known for Hitchcock-style cameos. As Carpenter departs the frame, the shadow of Father Malone, drinking wine as he trembles over the arrival of the pirates—he knows they are coming—is cast against the wall, this play with the arrangement of light and space a much more vivid inscription of Carpenter's authorship in the frame than his actual brief appearance as actor. Throughout *The Fog*, Carpenter's palpable presence as the storyteller—the one who invokes a ghostly presence through his literally invisible but nevertheless legible authorial hand, much as Machen dangles the pocket watch from offscreen in the film's opening shot—makes itself known via the presence of the fog, that cold dispersion that will drift to land and make the dead pirates terrifyingly corporeal. This fog, later in the film, will wind its way through screen space like a cloudy snake slithering around buildings and terrain; in one image, in which we see it approach a child and an old woman, the fog, moving slowly across the length of the frame, seems to be reaching out to hold the doomed house as if in a bodily gesture across the image. Earlier in the film, however, the fog, not quite yet wanting to make itself so corporeal and not yet fully on land, drifts invisibly through shots in this first nighttime sequence in the sleepy town, its lack of corporeality joining with Carpenter's crafty manipulation of the mise-en-scène. In a fashion perfectly befitting the Panavision frames it haunts, the first signal of the fog's arrival sends lateral ripples across the length of the image: a row of four pay phones, arranged diagonally across the screen, burst into sudden *ring-rings*, the coins inside them rattling; two circular mirrors, mounted in a convenience store, are placed on opposite sides of the wide frame, looking down at the store, with the right mirror jostling, as if about to fall, as if moved herky-jerky by an invisible hand; a convenience store worker goes about his after-hours sweeping and cleanup in an aisle as something made of glass falls to the floor and breaks offscreen; bottles and cans behind him and around him start to shake, fall, and shatter as this spirit of the fog, not quite yet visible, makes palpable its impending presence.

All of these mysteriously shaking and trembling objects seem to be shaking and trembling by themselves, the hands of the filmmakers that cause them to rattle kept carefully offscreen. Carpenter's arrangement of the objects and his careful masking of the hands that make them tremble situate the auteur as the visual correlative to the oral storyteller Machen from the opening sequence. Many of these same shots also imply Carpenter's sheer love for the palpable qualities of the Panavision frame, a pictorial taste here conjoined with the implication of the presence of the fog itself. The tendency, for example, of Panavision, when coupled with the wide-angle lenses Carpenter also favors, to bend or stretch the vertical lines of objects placed close to the lens, is a visual quality that marks several of the aforementioned shots, a subtle optical distortion that parallels the fog's own invisible control over mise-en-scène.

The fog, like Carpenter the artist, is in this way both an invisible and a palpably visible presence, manifest in impressions possible only in a widescreen format such as Panavision. The fog's tendency to stretch and bend itself across the frame—refusing to be fully contained in contingently narrowed views—is its most characteristic trait as a ghost, reminding us of Carpenter's own love for the bending and stretching of horizontal lines in his compositions. Other Carpenter creatures, as we have seen, refuse containment: The Shape, nearly always at the edge of the image; the Thing, its gruesomeness always already in excess of the wide frame that would seem the ideal canvas in which to fully depict it; and Christine, her sleek lines perfectly suited for Panavision but at the same time always prepared to race out of frame in murderous revved pursuit of quarry. But the fog is the only one of Carpenter's villains that seems to somehow instinctively know, to have as a part of its very ontological makeup, that it is meant to haunt Panavision, its lateral and slithering horizontal character, stretching leftward and rightward expansively like the frame itself, finding in its haunting of the frame its very and only purpose. "It wasn't meant for human beings. Just snakes," Fritz Lang once is reported to have said of CinemaScope, and Carpenter has taken the comment about snakes to heart in his depiction of the slithering fog in Panavision. In one scene, three boatmen spy the fog outside the window portals of their vessel. The portal takes on the shape of a circular iris through which we can see some of the fog, now palpably visible in the form of a cloudy blue line along the horizon line. A moment later it will take much more vivid, momentary, and grotesque corporeal shape as a pirate wielding a pickaxe, the weapon with which these boatmen are summarily and quickly dispatched. Characters throughout *The Fog* are frequently seen glimpsing the fog's bright blue dispersal across the frame through windows not unlike that little boat portal, windows that, like Carpenter's own frame, indicate the fog's presence without being able to fully contain it. When radio DJ Stevie Wayne (Adrienne Barbeau) becomes aware of the fog's impending arrival in the small town, spying it outside one of the film's many windows, she tries to warn her listeners, her acousmatic voice slipping and slithering like the fog across other images, as characters listen to her on their radios, her voice trying to become an aural protagonist suited to do battle with an equally dispersive villain. But the fog has her beat: as fog, it has the qualities of both visible palpability and invisible presence (like fog in real life, the fog in Carpenter's film just begs the viewer to reach out and touch it, albeit as we know, as with real fog, that we will not really be able to feel anything of much there). Slipping across the length and along the depth of Carpenter's images, this fog, drifting across these images long after the voice of Stevie has faded away from the frames, and long after the decay of notes from Carpenter's musical score have slipped away from our ears, always seems a few steps ahead of any sound that might try to catch up with it.

The Fog here contains another authorial mark prevalent to many Carpenter films: the director's own minimal score, written and performed on a synthesizer, evoking some of the minimalistic and dispersive qualities of this film's villain. Philip Brophy, in describing another Carpenter score—his music for *Escape from New York* (1981), his next film after *The Fog* and his second in his post-*Halloween* contract with Avco Embassy Pictures—writes words that are also apropos for an understanding of how Carpenter's music, one of his key touches as a film author, functions in *The Fog*. Brophy characterizes Carpenter's approach as "descoring," the use of minimalistic synthesizers to emphasize the holding of single notes to indicate tension, rather than the development of overtly complex or "melodramatic" orchestral music that might "mickey mouse" or anthropomorphically mimic the emotions and movements of the human characters:

> Played by synthesizer banks triggered by sequencers, the pulse is exceedingly inhuman. Metronomically controlled by voltage instead of wind-up mechanisms, tonally shaped from oscillators instead of acoustic materials, pitched by frequency controls instead of malleable tuning, the sound of music in *Escape from New York* is never actual. Rather, it is the by-product of electrical

> energy—always stated and presented as such and never allowed to "become music." . . . Positioned away from melodramatic statement, it is a narratological transposition of minimalism's drone and loop states, where the absence of horizontal melody intensifies one's awareness of the vertical depth in any one note or singular musical fragment. (99–100)

Carpenter's music parallels his handling of the Panavision frame, whereby widescreen becomes a visual transposition of the minimalistic qualities of sound. The film, like the music, eschews melodrama even in those moments in which the fog makes itself visibly palpable in the figural form of the undead pirates. In such moments the pirates are notable first and foremost for the way they are able to laterally arrange themselves across the frame, the way they *present* themselves in their poses to their soon-to-be-victims as screen figures, a presentation that relies on their ability to command the presence of the entire stretch of the image at least as much as it involves the sudden "swooping" of their deadly pickaxes from above the upper frame line, a gesture we see in the film's relatively few violent scenes. "The absence of the horizontal melody" in the music works in counterpoint to Carpenter's deft handling of the rhythm and modulation of his images, which, in his classical style, generate sequences that clearly and cleanly shape the narrative. Nevertheless, many of these images haunt us in ways that go beyond their functional place in a cleanly modulated, classical narrative, taking on, as so many images in Carpenter's cinema do, a distinctive status as haunting details that remain with us long after the details of the narrative drive have been forgotten. Such images attain a kind of "vertical depth," a singular fascination apart from their placement in any kind of narrative harmony, with textures that remain fascinating *as image*. I think of the glowing lights that appear under the crevices of doors and in the cracks of wood and window as the fog alights upon a house, near the end of the film, threatening a child and an old woman as, in its vivid state here as blue light, it becomes slowly something other than or in addition to fog. Such images serve the narrative even as they ultimately, in memory, slip away from it in their autonomy as images, freed from any obligation to permanent situation, much like fog itself.

In the Mouth of Madness

In the Mouth of Madness, which reunites Carpenter with cinematographer Gary B. Kibbe, tells the story of an insurance investigator, John Trent (Sam Neill), charged with deducing the whereabouts of infamous, reclusive, and suddenly missing horror novelist Sutter Cane (played with relish by Jürgen Prochnow when Cane is finally revealed about halfway through the movie). However, we are introduced to Cane initially not as a character but rather through the industrial, mechanical reproduction of his novels. The opening credits of *In the Mouth of Madness*, the main titles presented in cool, chilly blue, play against a series of images of a printing press in which Cane's recently completed novel, *The Hobb's*

End Horror, is being manufactured. Carpenter frames the printing press, across a series of shots of quick screen duration, so that the various positions of the camera create, in collaboration with the machinery itself, diverse arrangements of color, line, and volume, each image playing a part in a dynamically unfurling montage that finds surprising and varied artistry in the grinding repetition of mass reproduction. Unlike in the frames depicting the car factory in the opening scene of *Christine*, however, Carpenter does not show us any workers in these images helming the printing press here in *In the Mouth of Madness*. This book seems to be manufacturing itself, coming alive in these pulsing images as an autonomously sentient being. Perhaps Cane himself, that authorial spirit who will loom throughout the latter half of the film, is haunting this machinery, invisibly producing his text as object. In these delirious images in which a paperback is literally coming alive and taking on form and shape in a horror movie, Carpenter has already imbricated the authorship of his character Cane with his own palpable signature as a widescreen auteur.

Trent suspects Cane's sudden disappearance is a publicity stunt orchestrated by the author's publisher in order to whip up hysterical public interest in the new book, a theory the film never disproves and the possibility of which lingers behind its play with terrifying and surreal widescreen imagery. *In the Mouth of Madness* itself notes the way Cane's very name, and the styling of that name on the cover of his paperback editions, evokes Stephen King, a similarly "best-selling horror novelist who is so phenomenally popular that readers build their lives and their consciousness around his work, and wait with barely controlled patience for the appearance of his next novel" (Cumbow 204). In the film, readers of Cane's latest novel appear to lose their minds, as in the case of one thoroughly unhinged reader, an axe-wielding fellow who crashes through a bistro and threatens Trent early in the film. As characterized in Neill's performance, Trent initially appears too urbane and sophisticated to be duped by the illusions of mass-market genre confections—a cultivation and sense of self-confidence so extreme that he works to maintain it even after he is threatened with an axe by one of Cane's disturbed readers.

Trent's comfortable aloofness is sharply qualified by the fact that Trent himself, in the film's framing story, is ensconced in a mental institution and is relating the events of this tale to a psychologist (David Warner), the film's most overt reference to the narrative form of *The Cabinet of Dr. Caligari* (1920). Nevertheless, the meaning of the wide images of *In the Mouth of Madness* cannot be fully read simply as a flashback conveying some objective sense of what happened to Trent. As becomes clear, the events onscreen are apparently also being *written* by Cane, his hand as author joined with Carpenter's own in the crafting of these images. In this way, Trent himself, and all his psychological disturbances and pretensions to aloofness, is in fact situated as a character in a novel the diegetic author, like Carpenter the director, is toiling to finish. Trent, as a result, is not so much conveying the narrative of the film to us in flashback, as Nick

FIGURE 4.9 *In the Mouth of Madness* (New Line Cinema, 1994). Digital frame enlargement.

Holloway does throughout much of *Memoirs of an Invisible Man*. In that film, the wide image functions as a site toward which the invisible man's narration gestures, but also forms a precarious existential stage across which Nick must eventually move, from out of his comfortable and narrowed space of voice-over narration. Trent, by contrast, a figure not of his own narration but of Cane's devilish authorship, does not possess autonomous existential agency in the anamorphic frame of *In the Mouth of Madness*, even though Trent does behave throughout much of the film precisely in the belief that he possesses such control. Rather than the wide image serving as a stage on which Trent might confirm his heroism, he is instead, alternately delicately and devilishly, *held* by those frames. Consider, for example, the striking long shots of Trent in his padded cell in the framing story, as he begins to describe his terrors. All sense of authorial agency Trent might possess is thrown into counterpoint by how he is metaphorically held by the walls of his padded cell, as if they were outstretched hands, the walls' edges appearing in vertical curvature via Carpenter's use of the Panavision camera along with extreme wide-angle lenses (Figure 4.9). In his commentary track on the *In the Mouth of Madness* Blu-ray, Carpenter, as this sequence plays, conveys his passion for anamorphic framing, wide-angle lenses, and the fascinating optical distortions they can create. Where many designers of digital forms of 2.35:1 framing in the twenty-first century have worked to "correct" this edge-of-widescreen curvature in the design of lenses, presumably seeing in it only a "mistake," Carpenter prizes such traits for their unique aesthetic quality and distinction, seeing in them an optically specific type of visual beauty rather than something to be repressed under impersonal, corporate digital sheen. In this shot of Trent in the padded cell, the curved lines appear to hold the character gently, in delicate counterpoint to the mania of this unhinged asylum and the ensuing devilish ways in which Trent will be manipulated by the mastery of his author, Cane, later in the film.

Through this use of the anamorphic frame, *In the Mouth of Madness* stages a confrontation and occasional intersection between the hand of the filmmaker, Carpenter, who becomes intermittently, palpably present in effects such as the curved vertical lines in the wide-angle composition, and the hand of the author in the world of the story, Cane, whose authorial gestures as the generator of his forthcoming novel *In the Mouth of Madness* are implied parallels to the imagery of the film itself. However, Trent is unaware that he may be the puppet of a supreme author's manipulations. In his initial visit to Cane's publisher, Trent looks askance at the gothic book covers for Cane's various novels, framed and hung on the wall across the lavish main lobby. These covers are hung *safely* there, in this context becoming mere decor in an anodyne, corporate mise-en-scène (somewhat reminiscent of the corporate publishing house in Jean Negulesco's *The Best of Everything* at the outset of this book), not yet taking on their status as horrific works of art seeping into the consciousness of their readers in disturbing ways. But before showing us Trent looking at the book covers, Carpenter has flashed in front of us, in the form of a quick montage that is presumably also an initial sign of the effect this imagery is having on Trent, a series of gruesome illustrations from the cover illustrations of Cane's books, images that, when freed from their role simply as illustrations of whatever new narrative the publisher is peddling and put into the hands of an artist, become haunting monsters who engulf Carpenter's frame. Trent will soon later deduce the whereabouts of Cane by piecing together an arcane puzzle formed by slices of the covers of various Cane paperbacks, a visual clue that indicates Cane's Hobb's End, only ostensibly a fictional town, in fact is somewhere in New Hampshire.

Carpenter's authorship and Cane's authorship become increasingly imbricated as *In the Mouth of Madness* goes on. The film even ultimately suggests that Carpenter is himself a part of the world of his film. A figure named "John Carpenter," according to the theatrical one-sheet poster visible in a cinema Trent visits near the end of the film, directs the film adaptation of the Sutter Cane novel *In the Mouth of Madness* that Trent views in the final scene of the movie, after his encounter with Cane's text has apparently driven him completely mad.

We will work up to that screening, in which we discover that the version of *In the Mouth of Madness* Trent watches is filmed and projected as the very same anamorphic frame as the film we are presently watching. First, though, it is notable that the actual textual contents of Cane's novel, serving as the origin points for Carpenter's images of Trent's fall into a kind of vertigo of textuality, do occasionally become viewable in Carpenter's frame at key points. We glimpse a little bit of text in the printing press images that open the film, fragments of phrases and sentences telling us something about a character named "Carl" and his problems with his wife. (We will indeed meet doomed Carl and his murderous wife in a hotel in Hobb's End when Trent arrives there later in the movie.) The objective contents of the text are also invoked at various points in the film by Cane's editor, Linda Styles (Julie Carmen), the only one who has read the author's forthcoming *In the Mouth of Madness*. She seems to know what will happen before it happens once she and Trent reach Hobb's End, for Cane has already written it all in his book, a fiction in which Trent and Styles themselves now appear to be already written characters. Yet her mastery over Cane's authorship is limited in its omniscience, as Carpenter's widescreen frame becomes the site of Cane's own apparent intrusion into the visual form of *In the Mouth of Madness*—Styles has read only part of the novel because Cane is still busy finishing it. As Styles drives across a bridge leading into Hobb's End (Trent is asleep in the passenger seat), she looks down outside the car window and underneath the tires and edge of the car, which form a diagonal slant across the frame from the top left to the bottom right of the wide image. Below, she sees not the expected wood road of the bridge but rather a vast and vertiginous chasm of clouds and sky and thunder and lightning. We are disoriented not only by the surreal imagery of the frame of the floating sky under the car tires, an expansively vertiginous space, but also by the quick flashing of temporally distended imagery as the car journeys into Hobb's End. Carpenter's playfully manipulative hand as author, here, refiguring this ostensibly quotidian journey across a bridge, is joined with Cane's own, these images the cinematic correlative for the text of the novel *In the Mouth of Madness* he is presently still writing.

Carpenter's hand as author and Cane's as novelist will, however, be slightly disambiguated in those scenes in which Prochnow literally appears onscreen as Cane. In such sequences Cane himself, through Carpenter's direction and placement of the Prochnow who incarnates Cane in the frame, becomes ironically situated as a manipulated figure in a narrative that he elsewhere seems to control. Such situating also allows Carpenter the filmmaker to distance himself from the increasingly theological connotations of Cane's own authorship. After becoming aware of the surreal and horrific events taking place in Hobb's End, all of which we presume are transpositions and cinematic refigurations of events in the novel Cane is writing, Trent attempts to escape from the town. In doing so, he only manages to crash his car and is knocked unconscious. In the next shot, Trent is placed—by Cane, it is implied—in a priest's confessional. Carpenter frames this shot from above, looking down at Trent, who is helplessly held here much as he is in the padded cell in the film's framing story. In this shot, as in that earlier shot of the padded cell, Carpenter's anamorphic authorship is implicated in the curvature of the lines near the edges of the frame. And Trent again appears as a figure gently held by Carpenter's delicate wielding of lens and frame, even as the film through this technique also underscores the character's inability to inhabit that frame as conventional hero. Yet, unlike in the depiction of the framing story, Cane's presence becomes palpably figured in the manipulations of widescreen during the confessional sequence. Carpenter cuts from this bird's-eye-view, wide-angle shot to an eye-level shot of Trent in the confessional. The shot lacks the wide-angle curvature of vertical lines present in the previous shot, indicating that Carpenter, shifting here to a longer lens, is effacing one salient ait of his widescreen signature as the sequence progresses. On cue, Cane's own authorial signature becomes manifest in the frame. A bright light from the right side of the confessional appears, forming a diagonal expanse of light across the image. It becomes clear from the ensuing dialogue between Cane and Trent that Cane takes himself to be a kind of God in his authoring of popular paperbacks the fiction of which become incarnate in the frames of the film. Ensuing, extreme close-ups of both Trent and Cane focus us on the latter's malevolent authorial control, at times manifest in his ability to transport Trent like quicksilver from the site of a car accident to the interior of a confessional. This is, of course, something like Carpenter's ability, too, his control over the placement of Sam Neill and the joining of shots of the actor in the editing room creating a similar effect of transposition across frames. But for this moment Cane commands the frame of *In the Mouth of Madness* as the diegetic author of the frames we are presently watching. Whatever details or meaning we might find as we scan the wide image at this point in the film are implicitly claimed by Cane as originating from his authorial hand. Like Trent, we as viewers are becoming Cane's, and perhaps Carpenter's, puppets, scanning the image for meaning and yet falling vertiginously into predesigned artistry that mocks our pretensions to see in the wide image a source for free play with image and meaning.

In the next sequence we are suddenly transported to an image of Cane, at a typewriter, putting the finishing touches on his manuscript. Trent, now suddenly out of the confessional and splayed across a stone-tiled floor, can only look up at the commanding author who has just mastered his own text. The preceding scene in the confessional appears to have been the last scene Cane has written before finishing his book. *In the Mouth of Madness*, the novel, is complete. And yet now, in the film's most startling and memorable image, Cane reveals *himself* to be mere text. Slipping his fingers into the apparently pulpy surface of his face and the very fabric of the film *In the Mouth of Madness*, Cane pulls apart the anamorphic image of the movie that contains him, revealing underneath black type on white paper. These typed words are from Cane's novel, its fragments visible across these torn pages of the revealed text, ripped paper very like jagged teeth forming a kind of "mouth" leading into some unknown portal of terror. We had thought all along that the film we had been watching was in some way a representation of, or parallel to, the novel Cane had been writing, but Cane now reveals even that as a ruse, a play with surfaces obscuring even more devastating text lying underneath a wide frame whose expansiveness we have not even begun to understand. In doing so, Cane erases himself from the film, an erasure performed by Carpenter, the filmmaker, who now uses his frame to paint this picture of a novel so fulsomely evil it can manifest itself as a screen figure forming a portal into an otherworld. Now, our hero Trent is no longer manipulated by Cane but by Carpenter. We see Trent hesitantly step toward the rip in the fabric of the film formed by the torn pages of Cane's text. The camera then reverses position, and we see Trent from inside the very netherworld of horrific fiction across which he

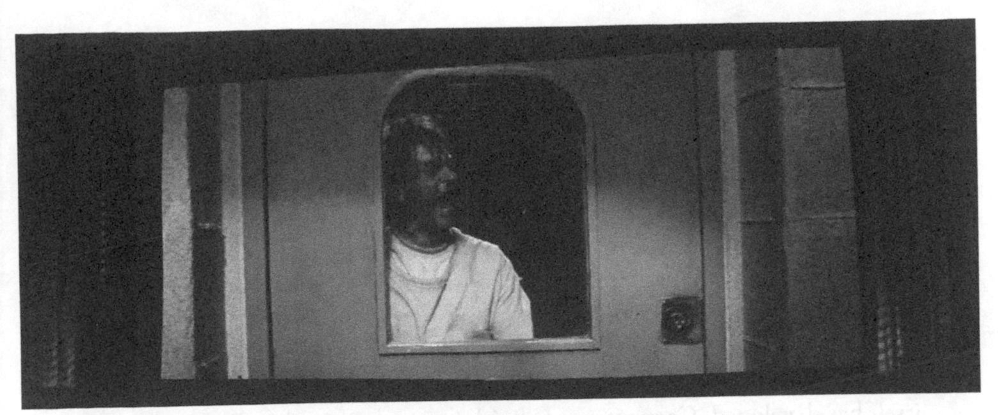

FIGURE 4.10 *In the Mouth of Madness* (New Line Cinema, 1994). Digital frame enlargement.

tentatively peers. As he does so, Styles reads words from the final pages of Cane's manuscript, implying that Cane is still authoring what we see. And perhaps that explanation makes sense for Trent and Styles within their fictional world. But for us, Cane's text has become part of the design of Carpenter's play with the wide image, his ability to visually express an appreciation for textual horror in the physical, cinematic world. Carpenter, not Cane, is the artist lingering behind these frames. And as if to clinch a claim to authorship superseding that of his fictional creation, Carpenter shows us, in the shots to follow, the creatures emerging from this netherworld to pursue Trent. These are delightfully rubbery monsters that do not fit with Cane's lofty, self-serious theological pretensions, and are playfully reflective of Carpenter's own affectionate love for 1950s horror and science fiction B-movies.

The final sequence of *In the Mouth of Madness* positions Trent, at least physically if not psychologically, very much as this book has sought to position an implied viewer of widescreen cinema, as one who sees in the space of the cinema itself ongoing possibilities of creative perception and engagement with expansive images. Psychologically, however, Trent can only laugh at any pretension to cinematic expanse when what he sees on the diegetic widescreen image is only his own demented image (Figure 4.10). Having been driven apparently mad by the commerce and currency of Cane's popular text, Trent escapes from the mental institution in the framing story (the vertiginous film now returning to whatever sense of "present time" we might be able to ascribe to it) and eventually makes his way to a movie theater showing the film adaptation of Cane's novel *In the Mouth of Madness*. The film, of course, stars John Trent, and is directed by John Carpenter, as the diegetic movie's one-sheet poster, framed outside the cinema, tells us. Something resembling an apocalypse appears to be occurring in the streets surrounding this cinema; Cane's intention to control the world through his prose has apparently become manifest. Meanwhile, inside, on the diegetic screen that takes the same shape as the anamorphic frame through which we have watched *In the Mouth of Madness*, we see events from earlier in the film unfurl, as Trent sits, with his bucket of popcorn, in thrall to the wide image.

Trent can no longer search for meaning but now can only be positioned to view a wide image that recollects, with absurd precision, his own demented experiences. But we presumably have not descended into madness in our own ongoing search for meaning in anamorphic imagery. There is a weirdly optimistic note struck here as Trent sits alone in this cinema and views shots from the film, not quite precisely an exact reproduction of the sequence of events in the film prior. Although an apocalypse would seem to consume the world of *In the Mouth of Madness*, Carpenter has already distanced himself from the fiction of Cane's theological terrors. And so, relatively autonomously from the narrative the film itself has developed, Carpenter's play with frames here in this final sequence of the film is ongoing, the diegetic screen becoming a site for a playful reorchestration of some of the frames from earlier in the movie. The moment, in counterpoint to the apocalyptic ending of the film, which might be metaphorically read as a "death of cinema" (or, specifically, a death of widescreen cinema), suggests an ongoing life for a cinema perceived in one way or another as wide—as beyond the normative frame—rather than conventionally narrow or repressive, even as the world burns all around Trent as he nibbles on his popcorn and gobbles up these images. In the final shot of the film, as Trent madly laughs at onscreen events that previously terrorized him, we might find even a little trace, should we want to see it, of the unpredictable, if perhaps now inescapably anxiety-ridden, pleasure still to be had in watching films that stretch widely across a wall, calling upon us to share in their delirious visions and possibilities.

And with that, the film ends.

Acknowledgments

I extend my deepest gratitude to the series editor of the Techniques of the Moving Image series, Murray Pomerance, for his gracious, encouraging, and thoughtful support of this project. I appreciate too his gift to me of Guy Davenport's book *Objects on a Table: Harmonious Array in Art and Literature,* which provided inspiration at a certain point in my work. The team at Rutgers University Press was, as always, immensely helpful, and I thank them for their time and careful attention to the manuscript. I also thank Nicole Solano for her steadfast editorial guidance of the book, which began with a vague idea about one director and ended as something much more interesting. The production and editing team at Rutgers University Press, particularly Susan Ecklund, brought their careful attention to the flow of both word and image across these pages, and for that I am very grateful.

Most of this book on widescreen cinema was written under ironic conditions, while in lockdown during a global pandemic than made watching expansive, wall-filling images on public movie screens sadly impossible. (It also made experimenting with some of the ideas at in-person academic conferences an impossibility, too, so I especially thank the peer reviewers of the manuscript for providing generous and constructive thoughts that helped improve the final version and sharpen both my thinking and my writing.) But before and after those lockdown months without cinema, screenings at repertory theaters, including screenings of a few films discussed in this book, helped sustain my love for film images projected with care by loving devotees of widescreen cinema as an art form meant for public, rather than private, exhibition. My idea for the book was sparked by a July 2019 screening of a beautifully preserved 35mm CinemaScope print of Howard Hawks's *The Land of the Pharaohs* at the Cinémathèque Française's "Plein les yeux!" ("Open Your Eyes!") series, a screening presided over by Jean Douchet, who would pass away only a few months later in November 2019. My modest travels, over the last several years, have taken me to several other fine repertory cinemas, including the Metrograph and the Quad in New York City, the Studio des Ursulines in Paris, the Plaza Theatre in Atlanta, and the Trylon Cinema in Minneapolis. On these various screens I have, at various times, been privileged to see some of the work of the directors examined in this book, in the form of the peerless beauty of their original 35mm celluloid incarnations projected on large screens. I thank the projectionists and programmers at these marvelous movie houses for their care and attention to the art of curation and exhibition, a practice still living with vibrancy through their labor, and which in no small part inspired many of the thoughts in this book.

As always, I treasure Jessica Belser for her companionship and love over the years.

Atlanta
Mankato
2023

Works Cited

Affron, Charles. *Cinema and Sentiment*. Chicago: University of Chicago Press, 1982.
Appelbaum, R. "From Cult Homage to Creative Control." *Films and Filming* 25 (1979): 9–16.
The Art of Blake Edwards: A Retrospective: Sculptures and Paintings 1969–2002. Exhibition catalog. West Hollywood: Pacific Design Center. 2009.
Banner, L. W. "The Creature from the Black Lagoon: Marilyn Monroe and Whiteness." *Cinema Journal* 47.4 (2008): 4–29.
Barr, Charles. "CinemaScope: Before and After." *Film Quarterly* 16.4 (1963): 4–24.
Bazin, André. "Massacre in CinemaScope." *André Bazin's New Media*. Ed. Dudley Andrew. Berkeley: University of California Press, 2014. 260–267.
———. "On Realism." *André Bazin: Selected Writings, 1943–1958*. Trans. Jacques Aumont. Montreal: Caboose, 2018. 5–8.
———. "Three Essays on Widescreen." Trans. Catherine Jones and Richard Neupert. *Velvet Light Trap* 21 (Summer 1985): 8–14.
Belton, John. *Widescreen Cinema*. Cambridge, MA: Harvard University Press, 1992.
Belton, John, Sheldon Hall, and Steve Neale. "Introduction." *Widescreen Worldwide*. Ed. John Belton, Sheldon Hall, and Steve Neale. New Barnet, UK: John Libbey, 2010. 1–4.
———. "Textual Analysis, Aesthetics, and Film Form." *Widescreen Worldwide*. Ed. John Belton, Sheldon Hall, and Steve Neale. New Barnet, UK: John Libbey, 2010. 59–61.
Berry, Sean. "Why Do We Watch Movies in Widescreen?" *Videomaker* December 26, 2018. <https://www.videomaker.com/news/why-do-we-watch-movies-in-widescreen/>.
Bordwell, David. *Poetics of Cinema*. New York: Routledge, 2008.
———. "Widescreen Aesthetics and Mise en Scene Criticism." *Velvet Light Trap* 21 (Summer 1985): 18–25.
Boulenger, Gilles. *John Carpenter: The Prince of Darkness*. Los Angeles: Silman-James, 2003.
Branigan, Edward. *Projecting a Camera: Language-Games in Film Theory*. London: Routledge, 2006.
Brophy, Philip. *100 Modern Soundtracks*. London: BFI, 2004.

Capua, Michelangelo. *Jean Negulesco: The Life and Films*. Jefferson: McFarland. 2017.
Cavell, Stanley. *Pursuits of Happiness: The Hollywood Comedy of Remarriage*. Cambridge, MA: Harvard University Press, 1981.
Chion, Michel. *The Voice in Cinema*. Trans. Claudia Gorbman. New York: Columbia University Press, 2008.
Conrich, Ian, and David Woods. "Introduction." *The Cinema of John Carpenter: The Technique of Terror*. Ed. Ian Conrich and David Woods. London: Wallflower, 2004. 1–9.
Cossar, Harper. *Letterboxed: The Evolution of Widescreen Cinema*. Lexington: University Press of Kentucky, 2011.
Cumbow, Robert C. *Order in the Universe: The Films of John Carpenter*. 2nd ed. Lanham: Scarecrow, 2000.
Davenport, Guy. *Objects on a Table: Harmonious Disarray in Art and Literature*. Washington: Counterpoint, 1998.
Deutelbaum, Marshall. "Basic Principles of Scope Composition." *Film History* 15.1 (2003): 72–80.
Dixon, Wheeler Winston. "The Multitrack World of *California Split*." *A Companion to Robert Altman*. Ed. Adrian Danks. Malden: Wiley-Blackwell, 2015. 166–183.
Dombrowski, Lisa. "Cheap but Wide: The Stylistic Exploitation of CinemaScope in Black-and-White Low-Budget American Films." *Widescreen Worldwide*. Ed. John Belton, Sheldon Hall, and Steve Neale. New Barnet, UK: John Libbey, 2010. 63–70.
Falwell, John. "The Art of Digression: Blake Edwards' *Skin Deep*." *Literature/Film Quarterly* 24.2 (1996): 177–182.
Ford, Hamish. "The Porous Frame: Visual Style in Altman's 1970s Films." *A Companion to Robert Altman*. Ed. Adrian Danks. Malden: Wiley-Blackwell, 2015. 119–145.
Fox, Jordan R. "Riding High on Horror (Interview with John Carpenter)." *Cinefantastique* 10.1 (Summer 1980): 5–10, 40, 42–44.
Gentry, Ric. "Fraker Records *Memoirs of an Invisible Man*." *American Cinematographer* 72.12 (1991): 46.
Gibbs, John. *The Life of Mise-en-Scène: Visual Style and British Film Criticism*. Manchester: Manchester University Press, 2013.
Gibbs, John, and Douglas Pye. "Preminger and Peckinpah: Seeing and Shaping Widescreen Worlds." *Widescreen Worldwide*. Ed. John Belton, Sheldon Hall, and Steve Neale. New Barnet, UK: John Libbey, 2010. 71–90.
Glitre, Kathrina. "Conspicuous Consumption: The Spectacle of Widescreen Comedy in the Populuxe Era." *Widescreen Worldwide*. Ed. John Belton, Sheldon Hall, and Steve Neale. New Barnet, UK: John Libbey 2010. 133–143.
Hall, Sheldon. "Carpenter's Widescreen Style." *The Cinema of John Carpenter: The Technique of Terror*. Ed. Ian Conrich and David Woods. London: Wallflower, 2004. 66–76.
Harvey, James. "Marilyn Reconsidered." *Threepenny Review* 58 (Summer 1994): 35–37.
Hass, Robert. *Twentieth Century Pleasures: Prose on Poetry*. New York: Norton, 2000.
Hilgers, Thomas. *Aesthetic Disinterestedness: Art, Experience, and the Self*. London: Routledge, 2017.
Homan, Catherine. "Whoever Cannot Give, Also Receives Nothing: Nietzsche's Playful Spectator." *The Philosophy of Play*. Ed. Emily Ryall, Wendy Russell, and Malcolm Maclean. London: Routledge, 2014. 98–108.
Houston, Penelope. "CinemaScope Productions." *Sight and Sound* 23.4 (1954): 198.
Howarth, Troy. *Assault on the System: The Nonconformist Cinema of John Carpenter*. Chicago: WK Books, 2020.

Jones, Catherine, and Richard Neupert. "Introduction." *Velvet Light Trap* 21 (Summer 1985): 1–7.
Jones, Kent. "John Carpenter: American Movie Classic." *Film Comment* 35.1 (1999): 26–31.
Keathley, Christian. *Cinephilia and History, or the Wind in the Trees*. Bloomington: Indiana University Press, 2006.
Kehr, Dave. "Blake Edwards." *The International Dictionary of Films and Filmmakers*. Chicago: St. James Press, 2001. 292–294.
Keller, Sarah. *Anxious Cinephilia: Pleasure and Peril at the Movies*. New York: Columbia University Press, 2020.
Kolker, Robert. *A Cinema of Loneliness: Penn, Stone, Kubrick, Scorsese, Spielberg, Altman*. 3rd ed. Oxford: Oxford University Press, 2000.
Konkle, Amanda. "How to (Marry a Woman Who Wants to) Marry a Millionaire." *Quarterly Review of Film and Video* 31.4 (2014): 364–383.
Lehman, Peter, and William Luhr. *Blake Edwards*. Athens: Ohio University Press. 1981.
———. *Blake Edwards: Returning to the Scene*. Athens: Ohio University Press. 1989.
Lethem, Jonathan. *They Live*. London: Soft Skull, 2011. Kindle version.
Liguoro, Francesca, and Giustina D'Oriano. "The Frontiers of Vision." *Le Cinémascope entre art et industrie*. Ed. Jean-Jacques Meusy. Paris: Association française de recherche sur l'historie du cinéma, 2003. 297–307.
Magee, Gayle. "Creativity and Compromise: *California Split*'s Original Soundtrack." *A Companion to Robert Altman*. Ed. Adrian Danks. Malden: Wiley-Blackwell, 2015. 210–230.
Mander, Jerry. *Four Arguments for the Elimination of Television*. New York: William Morrow, 1978.
Martin, Adrian. "Blake Edwards' Sad Songs of Love." Film Critic: Adrian Martin (website). Accessed April 6, 2023. <https://www.filmcritic.com.au/essays/edwards.html.> Originally published in *Freeze Frame* (July 1987): 10–13.
McElhaney, Joe. "*3 Women*: Floating Above the Awful Abyss." *A Companion to Robert Altman*. Ed. Adrian Danks. Malden: Wiley-Blackwell, 2015. 146–165.
Naremore, James. *Acting in the Cinema*. Berkeley: University of California Press, 1990.
Negulesco, Jean. "New Medium—New Methods." *New Screen Techniques*. Ed. M. Quigley Jr. New York: Quigley, 1953. 175–176.
Niemi, Robert. *The Cinema of Robert Altman: Hollywood Maverick*. London: Wallflower Press, 2016.
"Notes on *Mlle. Jeanne Samary*." Cincinnati Art Museum. Google Arts and Culture. Accessed April 29, 2022. <https://artsandculture.google.com/asset/mlle-jeanne-samary-pierre-auguste-renoir-french-b-1841-d-1919/WAFVxtONRd1ZXQ?hl=en.>
Peucker, Brigitte. *Incorporating Images: Film and the Rival Arts*. Princeton: Princeton University Press, 2014.
Pomerance, Murray. *Color It True: Impressions of Cinema*. New York: Bloomsbury, 2022.
———. "High Hollywood in *The Long Goodbye*." *A Companion to Robert Altman*. Ed. Adrian Danks. Malden: Wiley-Blackwell, 2015. 233–253.
Pratt, David. "Widescreen Box Office Performance to 1959." *Velvet Light Trap* 21 (Summer 1985): 65–66.
Price, Brian. "Color, the Formless, and Cinematic Eros." *Framework: The Journal of Cinema and Media* 47.1 (2006): 22–35.
Recuber, Tim. "Immersion Cinema: The Rationalization and Reenchantment of Cinematic Space." *Space and Culture* 10.3 (2007): 315–330.

Rivette, Jacques. "The Age of Metteurs en Scène." *Cahiers Du Cinéma, the 1950s: Neo-Realism, Hollywood, New Wave*. Ed. Jim Hillier. Cambridge, MA: Harvard University Press, 1985. 275–279.

Rogers, Ariel. *Cinematic Appeals: The Experience of New Movie Technologies*. New York: Columbia University Press, 2013.

Roggen, Sam. "CinemaScope and the Close-Up/Montage Style: New Solutions to Familiar Problems." *New Review of Film and Television Studies* 17:2 (May 2019). 148–184.

Rohmer, Éric. "The Cardinal Virtues of CinemaScope." *Cahiers Du Cinéma, the 1950s: Neo-Realism, Hollywood, New Wave*. Ed. Jim Hillier. Cambridge, MA: Harvard University Press, 1985. 281–283.

Ryall, Emily, Wendy Russell, and Malcolm Maclean, eds. *The Philosophy of Play*. London: Routledge, 2014.

Self, Robert T. "Art and Performance: Consolation at the End of Days." *Robert Altman: Critical Essays*. Ed. Rick Armstrong. Jefferson: McFarland, 2011. 156–179.

———. *Robert Altman's* McCabe & Mrs. Miller*: Reframing the American West*. Lawrence: University Press of Kansas, 2007.

———. *Robert Altman's Subliminal Reality*. Minneapolis: University of Minnesota Press, 2002.

Smith, Douglas. "'Up to Our Eyes in It': Theory and Practice of Widescreen in the French New Wave." *Studies in French Cinema* 17.2 (2017): 113–128.

Smith, Gavin and Richard T. Jameson. "The Movie You Saw Is the Movie We're Going to Make." *Robert Altman: Interviews*. Ed. David Sterrit. Jackson: University Press of Mississippi, 2000. 163–181.

Spellerberg, James. "CinemaScope and Ideology." *Velvet Light Trap* 21 (Summer 1985): 26–34.

Taylor, Giles. "A Military Use for Widescreen Cinema: Training the Body through Immersive Media." *Velvet Light Trap* 72 (Fall 2013): 17–32.

Thompson, David. *Altman on Altman*. London: Faber and Faber, 2011.

Truffaut, François. "A Full View." *Cahiers Du Cinéma, the 1950s: Neo-Realism, Hollywood, New Wave*. Ed. Jim Hillier. Cambridge, MA: Harvard University Press, 1985. 273–274.

Turvey, Malcolm. *Play Time: Jacques Tati and Comedic Modernism*. New York: Columbia University Press, 2020.

Varndell, Daniel. "Peter Sellers in *The Pink Panther*." *Close-Up, Great Cinematic Performances*. Vol. 1, *America*. Edinburgh: Edinburgh University Press. 2018. 135–146.

Warren, Charles. *Writ on Water: The Sources and Reach of Film Imagination*. Albany: State University of New York Press, 2022.

Wasser, Frederick. *Veni, Vidi, Video: The Hollywood Empire and the VCR*. Austin: University of Texas Press, 2002.

Wasson, Sam. *A Splurch in the Kisser: The Movies of Blake Edwards*. Middletown: Wesleyan University Press, 2010.

Wood, Robin. *Robin Wood on the Horror Film: Collected Essays and Reviews*. Ed. Barry Keith Grant. Detroit: Wayne State University Press, 2018.

Index

About the Author

STEVEN RYBIN is an associate professor of film studies at Minnesota State University, Mankato. He is the author of *Shots to the Heart: For the Love of Film Performance* (2022), *Geraldine Chaplin: The Gift of Film Performance* (2020), *Gestures of Love: Romancing Performance in Classical Hollywood Cinema* (2017), and *Michael Mann: Crime Auteur* (2013), among other books, edited volumes, book chapters, and articles.